THE SOLAR SYSTEM

GREAT ASTRONOMERS

THE SOLAR SYSTEM

GREAT ASTRONOMERS

SALEM PRESS
A Division of EBSCO Publishing
Ipswich, Massachusetts

GREY HOUSE PUBLISHING

Cover photo: Galileo Galilei by Ivan Petrovich Koler-Viliandi. (© The Gallery Collection/Corbis)

The Solar System: Great Astronomers, 2013, published by Grey House Publishing, Inc., Amenia, NY, under exclusive license from EBSCO Publishing, Inc.

∞ The paper used in these volumes conforms to the American National Standard for Permanence of Paper for Printed Library Materials, Z39.48 1992 (R1997).

paperback ISBN: 978-1-4298-3798-9
ebook ISBN: 978-1-4298-3804-7

PRINTED IN THE UNITED STATES OF AMERICA

CONTENTS

CONTRIBUTORS

Amy Ackerberg-Hastings

Bryan Aubrey

Raymond D. Benge, Jr.

Alvin K. Benson

Michael R. Bradley

Byron D. Cannon

Allan D. Charles

Patricia Cook

Thomas Drucker

George J. Flynn

Daniel J. Fuller

Sally Hibbin

James R. Hofmann

J. Donald Hughes

Wilbur R. Knorr

Paul E. Kuhl

R. C. Lutz

Yale R. Magrass

Timothy C. Miller

Bryan D. Ness

James Owen

Robert J. Paradowski

Francesca Rochberg-Halton

Elizabeth D. Schafer

Per Schelde

Roger Sensenbaugh

R. Baird Shuman

Amy Sisson

Ronald F. Smith

Katherine R. Sopka

Joseph L. Spradley

Brian Stableford

David L. Sterling

Rebecca Tolley-Stokes

William Urban

Brent Waters

Jane Carter Webb

Michael Witkoski

Frank Wu

Clifton K. Yearley

Kristen L. Zacharias

ABUL WEFA

Persian mathematician and astronomer

Abul Wefa played a major role in mathematics by developing sines and cosines as they apply to the field of trigonometry and used them to correct astronomical calculations carried forward from classical into Islamic times.

Born: June 10, 940; Buzjan, Khorasan (now in Iran)
Died: July 15, 998; Baghdad (now in Iraq)
Also known as: Abu-Wafa; Muhammad ibn Muhammad ibn Yahya ibn Isma il ibn al-Abbas
Abu al-Wafa al-Buzajani (full name)
Areas of achievement: Mathematics, astronomy, science and technology

Early Life

Born in 940, during the reign of the Abbasid caliph al-Mutaqqi, Abul Wefa (ah-BEWL weh-FAH) lived during a period of extraordinary cultural and intellectual productivity. His own fields of accomplishment, mathematics and astronomy, were already widely recognized as essential elements of high Islamic civilization. Very little seems to be known about Abul Wefa's early life. Apparently, his early education in mathematics occurred under the tutelage of two uncles, one of whom (Abu Amr al-Mughazili) had received formal training from the famous geometrician Abu Yahya al-Marwazi.

Whatever the possible source of patronage for the young man's further education may have been, his decision to move to Baghdad at the age of nineteen (in 959) greatly benefited the Abbasid court. Baghdad at this time was troubled politically, following the seizure of de facto control by a military clique headed by the Persian Buyid emirs; thereafter, the Buyids dominated the house of the caliphs until their fall from power in 1055. The Buyids were inclined to favor talented Persians who were drawn toward scholarly circles in the center of the empire. It is reported, for example, that it was Abul Wefa, himself then forty years of age and well established (c. 980), who introduced the Persian scholar and philosopher Abu Hayyan al-Tawhidi into the Baghdad entourage of the vizier Ibn Sa'dan. Abu Hayyan soon became famous under the vizier's patronage, composing a major work, *Al-Imta' wa'l mu'anasa* (a collection of notes drawn from philosophical and literary salon meetings), under a dedication to Ibn Sa'dan.

Patronage for Abul Wefa's work in courtly circles, however, must have come from a different milieu, that of the so-called Baghdad School. This scientific assembly flourished in the Abbasid capital in the last century before its conquest by the Seljuk Turks in 1055. According to some historians, patronage for the natural sciences in particular came precisely during the period in which Abul Wefa passed into the main stages of his scholarly career. The Buyid emir Adud al-Dawlah (r. 978-983) had nurtured an interest in astronomy through his own studies. He passed this interest on to his son, Saraf al-Dawlah, who built an observatory next to his palace and called scholars from all regions of the empire to glorify the reputation of his reign by carrying out scientific experiments. Abul Wefa was among this group.

Life's Work

The environment for learning in the Baghdad School, with its circle of eminent Islamic scientists, may explain how the young Persian scholar mastered so many technical fields in such a limited period of time. Beyond mere speculation regarding Abul Wefa's early personal contacts, however, one must consider the importance of translation work in the Baghdad School. Abul Wefa himself translated the work of the Greek algebraist Diophan-tus (fl. c. 250), who had explored the field of indeterminate algebraic equations. Abul Wefa was also known for his studies of, and commentaries on, Euclid. There are, however, no surviving texts to indicate what use he made of the work of these two forerunners from the classical pre-Islamic period.

By contrast, Abul Wefa's attention to the work of the second century Greek astronomer Ptolemy not only contributed to the preservation and transmission to the medieval West of the classical knowledge contained in Ptolemy's *Mathematike suntaxis* (c. 150; *Almagest*, 1952) but also earned for him an original and lasting reputation as an Islamic mathematician. The *Almagest* examined the field of trigonometry, which proposed mathematical relationships in terms of the angles and sides of right triangles. This called for the development of sines, or systematic relationships defined in a right triangle working from one of the acute angles, symbolically represented as A. Modern trigonometry expresses this relationship as $\sin A = a/c$, or $\sin A$ is equal to the ratio of the length of the side opposite that angle (a) to the length of the hypotenuse (c).

Ptolemy, in pioneering the field of spherical trigonometry, had laid down an approximate method for calculating sines (which he described as chords). Abul Wefa, however, drew on his studies of Indian precedents in the field of trigonometry that were unknown to Ptolemy, as well as models provided by Abul Wefa's predecessor al-Battani (858-929), to perfect Ptolemy's chords. This was done by applying algebraic, instead of geometric, methods of systematizing the sines. In particular, Abul Wefa's development of the half-chord made it possible to achieve much more precise measurements that would eventually be used in surveying and navigation. The most immediate application of his tables of sines, however, was in the field of astronomy.

One of Abul Wefa's contributions that left a legacy that lasted for many centuries involved the study of evection, or irregularity, in the longitude of the Moon. Later European commentators, including Louis Pierre E. A. Sédillot in the nineteenth century, looked at the Islamic astronomer's work and concluded that he, not Tycho Brahe (1546-1601), had been the first scientist to posit the theory of the "third inequality of the moon." Although this theory was later proved to be erroneous, the debate at least drew attention to the importance of Abul Wefa's originality in the field.

Abul Wefa himself compiled, in addition to his well-known tables of sines, a book of astronomical tables entitled *Zij al-wadih* (that which is clear). Like his earlier work on sines, this text is not extant in the original. Scholars tend to agree, however, that certain anonymous manuscripts preserved in European libraries, such as the *Zij al-shamil*, are taken from Abul Wefa's work.

Works that have survived and that have been at least partially translated include a book of arithmetic entitled Kita*b fi ma yahtaj ilayh al-kuttab wa l-'ummal min 'ilm al-hisab* (961-976; book on what is necessary from the science of arithmetic for scribes and businessmen), the *Kitab fi ma yahtaj ilayh al-sani 'min al-a'mal al-handasiyha* (after 990; book on what is necessary from geometric construction for the artisan), and a book entitled *Kitab al-kamil* (translated by Carra de Vaux in the *Journal Asiatique* of 1892). It is thought that Abul Wefa may have still been in Baghdad at the time of his death in 998.

Significance

Study of the Islamic cultural milieu in which Abul Wefa lived suggests a high level of syncretic interaction between ethnic subjects of the Baghdad caliphate—Arab, Persian, Greek, and other minorities. Abul Wefa's own career seems also to provide an example of a syncretic social hierarchy. Scientists and intellectual figures, it seems, had no reason to doubt that their accomplishments would be appreciated and supported by a ruling military elite whose social status was obviously determined by very different criteria. In this rather cosmopolitan period in Islamic history, there was room not only for scholars of diverse national origins at the caliph's court but also for representatives of different disciplines, secular and religious, to live side by side in a community that was truly representative of a world civilization. One can only understand the flourishing in Islam of such different disciplines (and the pure sciences in particular), however, if attention is given to the multiplicity of pre-Islamic sources that contributed both to the Baghdad caliphate itself and to the highly developed cultural institutions that it supported.

Byron D. Cannon

Further Reading

Bell, Eric T. *The Development of Mathematics*. New York: McGraw-Hill, 1945. Begins with a historical review of the field of mathematics from the first-known texts through successive stages of discoveries, ending at midpoint in the twentieth century. The chapter of most interest to students of Islamic science is entitled "Detour Through India, Arabia, and Spain, 400-1300." This title underscores the importance of the medieval period of Oriental history for the conservation of classical Western sources that only returned to Europe via the Islamic core zone, from eastern Iran to Spain.

Cajori, Florian. *A History of Mathematics*. 1931. 5th ed. Providence, R.I.: AMS Chelsea, 2000. This classic work has several important characteristics that still merit mention. It covers not only standard non-Western mathematical traditions (Hindu and Islamic) but also traditions from little-studied areas such as Mayan Central America and Japan. Includes detailed information on individual mathematicians' original findings.

Hogendijk, Jan P., and Abdelhamid I. Sabra, eds. *The Enterprise of Science in Islam: New Perspectives*. Cambridge, Mass.: MIT Press, 2003. A collection of essays surveying the history of Islamic science, including mathematics and astronomy. Illustrations, bibliography, index.

Huff, Toby E. *The Rise of Early Modern Science: Islam, China, and the West*. 2d ed. New York: Cambridge University Press, 2003. Provides a strong cross-cultural background for the rise of science in the Muslim world. Illustrations, bibliography, index.

Kennedy, Edward S. *Astronomy and Astrology in the Medieval Islamic World*. Brookfield, Vt.: Ashgate, 1998. Discusses astronomy, astronomers, math, and mathematicians, including Abul Wefa's calculation of the distance between Baghdad and Mecca. Index.

_______, ed. *Studies in the Islamic Exact Sciences*. Beirut: American University of Beirut Press, 1983. Provides a rather technical treatment of several scientific disciplines that flourished in early Islamic times, including the development, through trigonometry, of accurate astronomical calculations. A specific elaboration of Abul Wefa's work is included.

Nasr, Seyyed Hossein. *Islamic Science: An Illustrated Survey*. London: World of Islam Festival, 1976. A carefully researched photographic record of the tools of Islamic science. Textual treatment of historical figures such as Abul Wefa is more limited than in Nasr's text below. The choice of illustrations, however, particularly from Islamic astronomy, is so rich that the field itself becomes a much more coherent entity.

_______. *Science and Civilization in Islam*. Cambridge, Mass.: Harvard University Press, 1968. Because this work deals with the subject of science in Islamic civilization only, it can take time to explore individual contributions at some length.

ALHAZEN

Arab physicist, astronomer, and mathematician

Alhazen, Islam's greatest scientist, devoted his life to physics, astronomy, mathematics, and medicine. His treatise Optics, in which he deftly used experiments and advanced mathematics to understand the action of light, exerted a profound influence on many European natural philosophers.

Born: 965; Basra (now in Iraq)
Died: 1039; Cairo, Egypt
Also known as: Ibn al-Haytham; Abū ʻAlī al-hasan ibn al-Haytham (full name)

Early Life

Alhazen (ahl-HAH-zehn) was born in Basra. He was given a traditional Muslim education, but at an early age he became perplexed by the variety of religious beliefs and sects, because he was convinced of the unity of truth. When he was older, he concluded that truth could be attained only in doctrines whose matter was sensible and whose form was rational. He found such doctrines in the writings of Aristotle and in natural philosophy and mathematics.

By devoting himself completely to learning, Alhazen achieved fame as a scholar and was given a political post at Basra. In an attempt to obtain a better position, he claimed that he could construct a machine to regulate the flooding of the Nile. The Fātimid caliph al-Hākim (r. 996-1021?), wishing to use this sage's expertise, persuaded him to move to Cairo. Alhazen, to fulfill his boast, was trapped into heading an engineering mission to Egypt's southern border. On his way to Aswān, he began to have doubts about his plan, for he observed excellently designed and perfectly constructed buildings along the Nile, and he realized that his scheme, if it were possible, would have already been carried out by the creators of these impressive structures. His misgivings were confirmed when he discovered that the cataracts south of Aswān made flood control impossible. Convinced of the impracticability of his plan, and fearing the wrath of the eccentric and volatile caliph, Alhazen pretended to be mentally deranged; on his return to Cairo, he was confined to his house until al-Hākim's death around 1021.

Alhazen then took up residence in a small domed shrine near the al-Azhar Mosque. Having been given back his previously sequestered property, he resumed his activities as a writer and teacher. He may have earned his living by copying mathematical works, including Euclid's *Stoicheia* (c. fourth century B.C.E.; *Elements*, 1570) and Ptolemy's *Mathematike suntaxis* (c. 150; *Almagest*, 1952), and may also have traveled and had contact with other scholars.

Life's Work

The scope of Alhazen's work is impressive. He wrote studies on mathematics, physics, astronomy, and medicine, as well as commentaries on the writings of Aristotle and Galen. He was an exact observer, a skilled experimenter, and an insightful theoretician, and he put these abilities to excellent use in the field of optics. He has been called the most important figure in optics between antiquity and the seventeenth century. Within optics itself, the range of his interests was wide: He discussed theories of light and vision, the anatomy and diseases of the eye, reflection and refraction, the rainbow, lenses, spherical and parabolic mirrors, and the pinhole camera (camera obscura).

Alhazen's most important work was *Kitāb al-Manāzir*, commonly known as *Optics*. Published first in Latin

(*Opticae thesaurus Alhazeni libri vii*, 1572) and partially in English as *The Optics of Ibn al-Haytham: Books I-III, On Direct Vision* (1989), it attempted to clarify the subject by inquiring into its principles. He rejected Euclid's and Ptolemy's doctrine of visual rays (the extramission theory, which regarded vision as analogous to the sense of touch). For example, Ptolemy attributed sight to the action of visual rays issuing cortically from the observer's eye and being reflected from various objects. Alhazen also disagreed with past versions of the intromission theory, which treated the visible object as a source from which forms (simulacra) issued. The atomists, for example, held that objects shed sets of atoms as a snake sheds its skin; when a set enters the eye, vision occurs. In another version of the intromission theory, Aristotle treated the visible object as a modifier of the medium between the object and the eye.

Alhazen found the atomistic theory unconvincing because it could not explain how the image of a large mountain could enter the small pupil of the eye. He did not like the Aristotelian theory because it could not explain how the eye could distinguish individual parts of the seen world, since objects altered the entire intervening medium. Alhazen, in his version of the intromission theory, treated the visible object as a collection of small areas, each of which sends forth its own ray. He believed that vision takes place through light rays reflected from every point on an object's surface converging toward an apex in the eye.

According to Alhazen, light is an essential form in self-luminous bodies, such as the sun, and an accidental form in bodies that derive their luminosity from outside sources. Accidental light, such as the moon, is weaker than essential light, but both forms are emitted by their respective sources in exactly the same way: noninstantaneously, from every point on the source, in all directions, and along straight lines. To establish rectilinear propagation for essential, accidental, reflected, and refracted radiation, Alhazen performed many experiments with dark chambers, pinhole cameras, sighting tubes, and strings.

In the first book of *Optics*, Alhazen describes the anatomy of the eye. His description is not original, being based largely on the work of Galen, but he modifies traditional ocular geometry to suit his own explanation of vision. For example, he claims that sight occurs in the eye by means of the glacial humor (what would be called the crystalline lens), because when this humor is injured, vision is destroyed. He also uses such observations as eye pain while gazing on intense light and afterimages from strongly illuminated objects to argue against the visual-ray theory, because these observations show that light is coming to the eye from the object. With this picture of intromission established, Alhazen faces the problem of explaining how replicas as big as a mountain can pass through the tiny pupil into the eye.

He begins the solution of this problem by recognizing that every point in the eye receives a ray from every point in the visual field. The difficulty with this punctiform analysis is that, if each point on the object sends light and color in every direction to each point of the eye, then all this radiation would arrive at the eye in total confusion; for example, colors would arrive mixed. Simply put, the problem is a superfluity of rays. To explain vision, each point of the surface of the glacial humor needs to receive a ray from only one point in the visual field. In short, it is necessary to establish a one-to-one correspondence between points in the visual field and points in the eye.

To fulfill this goal, Alhazen notices that only one ray from each point in the visual field falls perpendicularly on the convex surface of the eye. He then proposes that all other rays, those falling at oblique angles to the eye's surface, are refracted and so weakened that they are incapable of affecting visual power. Alhazen even performed an experiment to show that perpendicular rays are strong and oblique rays weak: He shot a metal sphere against a dish both perpendicularly and obliquely. The perpendicular shot fractured the plate, whereas the oblique shot bounced off harmlessly. Thus, in his theory, the cone of perpendicular rays coming into the eye accounts for the perception of the visible object's shape and the laws of perspective.

Book 2 of *Optics* contains Alhazen's theory of cognition based on visual perception, and book 3 deals with binocular vision and visual errors. Catoptrics (the theory of reflected light) is the subject of book 4. Alhazen here formulates the laws of reflection: Incident and reflected rays are in the same plane, and incident and reflected angles are equal. The equality of the angles of incidence and reflection allows Alhazen to explain the formation of an image in a plane mirror. As throughout *Optics*, Alhazen here uses experiments to help establish his contentions. For example, by throwing an iron sphere against a metal mirror at an oblique angle, he found that the incident and reflected movements of the sphere were symmetrical. The reflected movement of the iron sphere, because of its heaviness, did not continue in a straight line, as the light ray does, but Alhazen did not contend that the iron sphere is an exact duplicate of the light ray.

Alhazen's investigation of reflection continues in books 5 and 6 of *Optics*. Book 5 contains the famous "Problem of Alhazen": For any two points opposite a spherical reflecting surface, either convex or concave, find the point or points on the surface at which the light from one of the two points will be reflected to the other. Today it is known that the algebraic solution of this problem leads to an equation of the fourth degree, but Alhazen solved it geometrically by the intersection of a circle and a hyperbola.

Book 7, which concludes *Optics*, is devoted to dioptrics (the theory of refraction). Although Alhazen did not discover the mathematical relationship between the angles of incidence and refraction, his treatment of the phenomenon was the most extensive and enlightening before that of René Descartes. As with reflection, Alhazen explores refraction through a mechanical analogy. Light, he says, moves with great speed in a transparent medium such as air and with slower speed in a dense body such as glass or water. The slower speed of the light ray in the denser medium is the result of the greater resistance it encounters, but this resistance is not strong enough to hinder its movement completely. Because the refracted light ray is not strong enough to maintain its original direction in the denser medium, it moves in another direction along which its passage will be easier (that is, it turns toward the normal). This idea of the easier and quicker path was the basis of Alhazen's explanation of refraction, and it is a forerunner of the principle of least time associated with the name of Pierre de Fermat.

This frontispiece to Alhazen's Opticae thesaurus *depicts various uses of optics, including Archimedes' defense of Syracuse with mirrors.* (Library of Congress)

Optics was Alhazen's most significant work and by far his best known, but he also wrote more modest treatises in which he discussed the rainbow, shadows, camera obscura, and Ptolemy's optics as well as spheroidal and paraboloidal burning mirrors. The ancient Greeks had a good understanding of plane mirrors, but Alhazen developed an exhaustive geometrical analysis of the more difficult problem of the formation of images in spheroidal and paraboloidal mirrors.

Although Alhazen's achievements in astronomy do not equal those in optics, his extant works reveal his mastery of the techniques of Ptolemaic astronomy. These works are mostly short tracts on minor problems, for example, sundials, moonlight, eclipses, parallax, and determining the *gibla* (the direction to be faced in prayer). In another treatise, he was able to explain the apparent increase in size of heavenly bodies near the horizon, and he also estimated the thickness of the atmosphere.

His best astronomical work, and the only one known to the medieval West, was *Hay'at al-'alan* (tenth or eleventh century; *Ibn al-Haytham's On the Configuration of the World* 1990). This treatise grew out of Alhazen's desire that the astronomical system should correspond to the true movements of actual heavenly bodies. He therefore attacked Ptolemy's system, in which the motions of heavenly bodies were explained in terms of imaginary points moving on imaginary circles. In his work, Alhazen tried to discover the physical reality underlying Ptolemy's abstract astronomical system. He accomplished this task by viewing the heavens as a series of concentric spherical shells whose rotations were interconnected. Alhazen's system accounted for the apparent motions of the heavenly bodies in a clear and nontechnical way, which accounts for the book's popularity in the Middle Ages.

Alhazen's fame as a mathematician has largely depended on his geometrical solutions of various optical problems, but more than twenty strictly mathematical treatises have survived. Some of these deal with geometrical problems arising from his studies of Euclid's *Elements*, whereas others deal with quadrature problems, that is, constructing squares equal in area to various plane figures. He also wrote a work on lunes (figures contained between the arcs of two circles) and on the properties of conic sections. Although he was not successful with every problem, his performance, which exhibited his masterful command of higher mathematics, has rightly won for him the admiration of later mathematicians.

Significance

Alhazen was undoubtedly the greatest Muslim scientist, and *Optics* was the most important work in the field from Ptolemy's time to the time of Johannes Kepler. He extricated himself from the limitations of such earlier theories as the atomistic, Aristotelian, and Ptolemaic and integrated what he knew about medicine, physics, and mathematics into a single comprehensive theory of light and vision. Although his theory contained ideas from older theories, he combined these ideas with his new insights into a fresh creation, which became the source of a new optical tradition.

His optical theories had some influence on Islamic scientists, but their main impact was on the West. *Optics* was translated from Arabic into Latin at the end of the twelfth century. It was widely studied, and in the thirteenth century, Witelo (also known as Vitellio) made liberal use of Alhazen's text in writing his comprehensive book on optics. Roger Bacon, John Pecham, and Giambattista della Porta are only some of the many thinkers who were influenced by Alhazen's work. Indeed, it was not until Kepler, six centuries later, that work on optics progressed beyond the point to which Alhazen had brought it. Even Kepler, however, used some of Alhazen's ideas, for example, the one-to-one correspondence between points on the object and points in the eye. It would not be going too far to say that Alhazen's optical theories defined the scope and goals of the field from his day to today.

Robert J. Paradowski

Further Reading

Bakar, Osman. *The History and Philosophy of Islamic Science*. Cambridge, England: Islamic Texts Society, 1999. Discusses questions of methodology, doubt, spirituality and scientific knowledge, the philosophy of Islamic medicine, and how Islamic science influenced medieval Christian views of the natural world.

Grant, Edward, ed. *A Source Book in Medieval Science*. Cambridge, Mass.: Harvard University Press, 1974. A compilation of readings from medieval natural philosophers, including several selections in English translation from the works of Alhazen. Bibliography, index.

Hayes, John R., ed. *The Genius of Arab Civilization: Source of Renaissance*. 3d ed. New York: New York University Press, 1992. In this beautifully illustrated book, several international authorities discuss the intellectual achievements of Islamic culture. Sabra's chapter on the exact sciences contains an account of Alhazen's work in the context of Islamic intellectual history. Bibliography, index.

Hogendijk, Jan P., and Abdelhamid I. Sabra, eds. *The Enterprise of Science in Islam: New Perspectives*. Cambridge, Mass.: MIT Press, 2003. A collection surveying the history of science, including mathematics, optics, and astronomy, in Islam, with a chapter on Alhazen's work. Illustrations, bibliography, index.

Huff, Toby E. *The Rise of Early Modern Science: Islam, China, and the West*. 2d ed. New York: Cambridge University Press, 2003. Provides a strong cross-cultural background for the rise of science and medicine in the Muslim world. Illustrations, bibliography, index.

Lindberg, David C. *Theories of Vision from al-Kindi to Kepler*. Chicago: University of Chicago Press, 1981. Surveys visual theory against the background of ancient accomplishments. The chapter on Alhazen's intromission theory is excellent. Bibliography, index.

_______, ed. *Science in the Middle Ages*. Chicago: University of Chicago Press, 1978. Through the expertise of several historians of medieval science, this book examines in depth all major aspects of natural philosophy in the Middle Ages. The approach is not encyclopedic but interpretative. Includes a chapter on optics, in which Alhazen's work is clearly explained.

Nasr, Seyyed Hossein. *Islamic Science: An Illustrated Study*. Westerham, England: World of Islam Festival,

1976. The first illustrated study ever undertaken of the whole of Islamic science. Using traditional Islamic concepts, the author discusses various branches of science, including optics.

_______. *Science and Civilization in Islam.* Cambridge, Mass.: Harvard University Press, 1968. This book is the first one-volume work in English to deal with Islamic science from the Muslim rather than the Western viewpoint. Its approach is encyclopedic rather than analytic, but it does contain a discussion of Alhazen's work in its Muslim context.

Sabra, Abdelhamid I. *Optics, Astronomy, and Logic: Studies in Arabic Science and Philosophy.* Brookfield, Vt.: Variorum, 1994. A history of science, specifically optics and astronomy, in the Muslim world of Alhazen's time. Includes discussion of Alhazen and his work. Illustrations, index.

_______. *Theories of Light from Descartes to Newton.* New York: Cambridge University Press, 1981. Though this book is mainly centered on seventeenth century theories of light, the author discusses in detail the impact of Alhazen's ideas on the optical discoveries of such thinkers as Descartes and Christiaan Huygens. Bibliography, index.

ANAXIMANDER

Greek philosopher, astronomer, and cartographer

Anaximander invented the scientific use of models and maps. He also realized, contrary to the prevailing thought of his day, that the original substance of matter must be an eternal, unlimited reservoir of qualities and change.

Born: c. 610 B.C.E.; Miletus, Greek Asia Minor (now in Turkey)
Died: c. 547 B.C.E.; probably Miletus
Areas of achievement: Natural history, astronomy, geography

Early Life

Anaximander (uh-nak-suh-MAN-duhr) was a fellow citizen and student of Thales, the Milesian usually credited with having inaugurated Western philosophy. Thales, some forty years older than his protégé, put none of his philosophical thought in writing and maintained no formal pedagogical associations with pupils. However, Thales' cosmological views (as reconstructed by historians) doubtless inspired Anaximander, and Anaximander finally expanded on Thales' ideas with innovative leaps in conceptual abstraction.

Anaximander was known in his day for his practical achievements and his astronomical discoveries. He is said to have been chosen by the Milesians as the leader for a new colony in Apollonia on the Black Sea. He traveled widely and was the first Greek to publish a "geographical tablet," a map of the world. The map was circular, and it was centered on the city of Delphi, because Delphi was the location of the *omphalos*, or "navel" stone, that was thought to be the center of Earth. Anaximander is also said to have designed a celestial map and to have specified the proportions of stellar orbits. In addition to the celestial map, he built a spherical model of the stars and planets, with Earth located at the center and represented as a disk or cylinder whose height was one third its diameter. The heavenly bodies were rings of hollow pipe of different sizes that were placed on circling wheels in ratios of three to six to nine, in proportion to the magnitude of Earth. This model was dynamic; the wheels could be moved at different speeds, making it possible to visualize patterns of planetary motion. Anaximander is also credited with inventing the gnomon (part of a sundial) and with having discovered the zodiac.

All these eclectic interests and discoveries illustrate, with elegance, Anaximander's particular genius, namely, his rational view of the world. This way of thinking was quite an innovation at a time when both scientific and protophilosophical thought took their content from the mythical and literary traditions, and thus were marked by vagueness and mystery. Anaximander viewed the world as steadily legible; he had the expectation of its rational intelligibility. His map of the world and his model of the heavens show his anticipation of symmetry and order. Earth, he argued, remained at rest in the center of the cosmos by reason of its equidistance at all points to the celestial circumference; it had no reason to be pulled in one direction in preference to any other. He projected the celestial orbits in perfect and pleasing proportions, and he anticipated regular motions.

Anaximander's mapping and modeling techniques themselves were products of his rationalistic thinking. Models and maps relocate some set of unified phenomena into a new level of abstraction. Implicit in map and model design is the assumption that the abstractions will preserve the intelligible relationships present in the world that they reproduce. Thus Anaximander's introduction of models and maps represents a tremendous and utterly original conceptual leap from the world "seen" to the world's operations understood and faithfully reproduced by the abstracting human mind.

Life's Work

Anaximander's rational view of the world received its fullest and most innovative expression in his philosophy of nature. Here one finds the first unified and all-encompassing picture of the world of human experience in history that is based on rational deduction and explanation of all phenomena.

In order to understand Anaximander properly, his terminology must be put into its historical context. What Anaximander (and Thales as well) understood by "nature" is not quite the same as its modern sense. In Ionian Greece, *physis* denoted the process of growth and emergence. It also denoted something's origin, or source, that from which the thing is constantly renewed. Nature, in the Ionian sense of *physis*, had nothing to do with matter; even Aristotle was mistaken in thinking that it did. In fact, no word for matter even existed in Anaximander's day. It is also important to note that Anaximander's thought is reconstructed entirely from ancient secondary sources. The one extant fragment of Anaximander's own words is the quotation of an ancient historian. Thus, any explication of Anaximander's thought is, to some extent, conjectural and interpretive.

Anaximander's philosophy of nature arose in part as a response to Thales' ideas on nature. Thales held that water was the nature of everything. This meant, in the light of the ancient idea of *physis*, that water was the origin of everything, that everything was sustained by, and constantly renewed from, water. This notion does not have any allegorical or mythical connotations in Thales' formulation. Water is the ordinary physical stuff in the world, not some engendering god such as the Oceanus of Thales' predecessors. That is the reason Thales is the first philosopher: He had a theory about the origin of things that competed with ancient creation myths.

Anaximander agreed with Thales that the origin of the things of the world was some common stuff, but he thought that the stuff could not be some ordinary element. He rejected Thales' conception on purely logical grounds, and his reasoning was quite interesting. How could any manifestly singular stuff ever give rise to qualities that pertained to things differently constituted, such as earth and fire? What is more, if water were the source of things, would not drying destroy them? Thus, reasoned Anaximander, the thing with which the world begins cannot be identical with any of the ordinary stuff with which humans are acquainted, but it must be capable of giving rise to the wide multiplicity of things and their pairs of contrary qualities. What therefore distinguishes the source from the world is that the source itself is "unbounded": It can have no definite shape or quality of its own but must be a reservoir from which every sort or characteristic in the world may be spawned. So Anaximander called the source of things this very name: *apeiron*, Boundlessness, or the Boundless. Anaximander designated the Boundless an *arche*, a beginning, but he did not mean a temporal beginning. The Boundless can have no beginning, nor can it pass away, for it can have no bounds, including temporal ones.

Thus the eternal source, the Boundless, functions as a storehouse of the world's qualities, such that the qualities that constitute some present state of the world have been separated out of the stock, and when their contrary qualities become manifest, they will, in turn, be reabsorbed into the reservoir. When Earth is hot, heat will come forth from the Boundless; when Earth cools, cold will come forth and heat will go back. For Anaximander, this process continued in never-ending cycles.

The cause of the alternating manifestations of contrary qualities is the subject of the single existing fragment of Anaximander's own words, the only remains of the first philosophy ever written. Out of the Boundless, Anaximander explains, the worlds arise, but from whatever things is the genesis of the things that are, into these they must pass away according to necessity; for they must pay the penalty and make atonement to one another for their injustice according to the ordering of Time.

History has produced no consensus of interpretation for this passage and its picturesque philosophical metaphor for the rationale of the world. Anaximander was probably thinking of a courtroom image. Each existing thing is in a state of "having-too-much," so that during the time it exists it "commits injustice" against its opposite by preventing it from existing. In retribution, the existing thing must cede its overt existence for its opposite to enjoy and pay the penalty of returning to the submerged place in the great Boundless reservoir. This cycling, he added, is how Time is ordered or measured. Time is the change, the alternating manifestation of opposites.

Here is the apotheosis of Anaximander's rational worldview. The world's workings are not simply visible and perspicuous, but neither are they whimsical and mysterious. The hidden workings of things may be revealed in the abstractions of the human mind. The world works, and is the way that it is, according to an eternal and intelligible principle. What is more, this world and its workings are unified, indeed form a cosmos. The cosmos, in turn, can be understood and explained by analogy with the human world; the justice sought in the city's courts is the same justice that sustains everything that human perception finds in the universe.

Significance

Classical antiquity credited Thales with having pioneered philosophy. Anaximander, with his scientific curiosity and his genius for abstract insight, poised philosophical inquiry for new vistas of exploration; his new philosophical approach inaugurated penetrating, objective analysis. His principle of the eternal Boundless as the source of the world's multifarious qualities and change forms the conceptual backdrop against which twenty-five centuries of science and natural philosophy have developed.

Two particular innovations of Anaximander have never been abandoned. First, his extension of the concept of law from human society to the physical world continues to dictate the scientific worldview. The received view in Anaximander's time—that nature was capricious and anarchic—has never again taken hold. Second, Anaximander's invention of the use of models and maps revolutionized science and navigation and continues to be indispensable, even in people's daily lives. All scientific experiments are models of a sort: They are laboratory-scale contrivances of events or circumstances in the world at large. Purely visual three-dimensional models continue to be crucial in scientific discoveries: the so-called Bohr model of the atom played a crucial role in physics; the double-helical model was important to the discovery of the structure and function of deoxyribonucleic acid (DNA). Maps are taken for granted now, but if human beings had relied on verbal descriptions of spatial localities, civilization would not have proceeded very far.

Thus, Anaximander's innovations and influence persist. Indeed, it is difficult to imagine a world without his contributions. Anaximander himself could hardly have seen all the implications of his discoveries, for even now one can only guess at the future direction of abstract thought.

Patricia Cook

Further Reading

Brumbaugh, Robert S. *The Philosophers of Greece*. Albany: State University of New York Press, 1981. This volume contains a short, digested chapter on Anaximander's life and accomplishments. Emphasizes cartography and engineering. Includes a reproduction of the first map designed by Anaximander.

Burnet, John. *Early Greek Philosophy*. 4th ed. New York: Meridian Books/World, 1961. A detailed scholarly analysis of Anaximander's thought in the context of comparisons with, and influences on, other pre-Socratic philosophers.

Couprie, Dirk L., Robert Hahn, and Gerard Naddaf. *Anaximander in Context: New Studies in the Origins of Greek Philosophy*. Albany: State University of New York Press, 2002. Examines the social, political, cosmological, astronomical, and technological backgrounds in which Anaximander's thought developed.

Guthrie, W. K. C. *The Earlier Presocratics and the Pythagoreans*. Vol. 1 in *A History of Greek Philosophy*. New York: Cambridge University Press, 1978-1990. Contains a chapter on Anaximander's cosmology. Focuses in a very close analysis on the concepts of *apeiron* and *apeiron* as *arche*.

Hahn, Robert. *Anaximander and the Architects: The Contributions of Egyptian and Greek Architectural Technologies to the Origins of Greek Philosophy*. Albany: State University of New York Press, 2001. Posits that technologies emerging among Egyptian and Greek architects strongly impacted the development of Greek philosophy.

Kahn, Charles H. *Anaximander and the Origins of Greek Cosmology*. Indianapolis: Hackett, 1994. Surveys the documentary evidence for Anaximander's views, reconstructs a detailed cosmology from documentary texts, and devotes an entire chapter to analysis and interpretation of Anaximander's fragment.

Kirk, Geoffrey S., and John E. Raven. *The Presocratic Philosophers*. 2d ed. New York: Cambridge University Press, 1983. Contains a chapter on Anaximander and a close formal analysis of textual testimony on Anaximander's thought.

Wheelwright, Philip, ed. *The Presocratics*. New York: Odyssey Press, 1966. A primary source. Contains the Anaximander fragment in translation. Also contains testimonies from Aristotle and other Greek and Latin sources who read and commented on Anaximander's treatise.

ĀRYABHATA THE ELDER

Indian mathematician and astronomer

Āryabhata the Elder's treatise, The Aryabhatiya, is the first work of Indian mathematics that has a definite author and date. It indicates what Indian mathematicians had accomplished by the end of the fifth century.

Born: c. 476; possibly Ashmaka or Kusumapura (now in India)
Died: c. 550; place unknown
Also known as: Āryabhata I; Āryabhatta

Early Life

Āryabhata (ahr-yah-BAH-tuh) the Elder is a figure of whom almost nothing is known other than his work. His only surviving book, the *Aryabhatiya* (499; *The Aryabhatiya*, 1927), indicates that it was written in 499 and that he was twenty-three years old at the time, which provides the traditional date for his birth. There is a good deal of disagreement about the exact date, as well as about the place of his birth. One of his successors in the field of Indian mathematics refers to Āryabhata as being from Ashmaka, a region in the northwestern part of modern India. There was a migration from that region to a more southerly part of India, and Āryabhata is likely to have been born there, as it is close to the only geographic region in which he is known to have worked. He is called Āryabhata the Elder to distinguish him from a later mathematician who wrote under the same name. The later Āryabhata may have chosen to use that as the signature for his works as a tribute to his predecessor.

In the course of the mathematical section of *The Aryabhatiya*, the author refers to the town of Kusumapura. Most scholars believe that this refers to the town where he was doing his academic work rather than his birthplace, but others argue that this town was where the mathematician was born. Kusumapura is frequently identified as the modern-day Patna, the capital of the region of Bihar. Because his work was to prove influential in the Kerala school of astronomy, practiced at the very south tip of India, some scholars have claimed that he spent his career there, but evidence to support that claim is no easier to come by than evidence regarding any other feature of Āryabhata's life.

The nature of the education that Āryabhata received is unknown, but from the text of his work it is clear that he studied Greek mathematics and astronomy as well as the Indian traditions in both areas. Because Āryabhata's work is the first systematic treatment of these subjects that has survived from that part of the world, it may be that he was dependent on odd scraps of knowledge from his predecessors. He would have had plenty of examples of particular sorts of calculation from which to learn. Because he wrote in Sanskrit verse, his training would have been literary as well as scientific. From the religious tone of his invocations, it seems safe to conclude that he was not a Buddhist, as he pays reverence to Brahman and the astronomical bodies.

Life's Work

From later references, it appears that Āryabhata wrote a book on astronomy as well as *The Aryabhatiya*. There were periods of time when neither work was available to students, but *The Aryabhatiya* was subsequently made generally available. The work on astronomy is not known to have survived.

The Aryabhatiya is a Sanskrit poem divided into four sections: an introduction, a mathematical section, a section on time measurement, and a section on celestial spheres. The opening section is written in a different verse form than the others, as though it may have been more like an invocation. There is an invocation at the start of the section on mathematics as well. The entire mathematical section is only thirty-three verses long, which attests to the brevity of his style. As a result, there is no illustration of the techniques and ideas he describes by numerical examples.

By Āryabhata's time, the use of numerals that were the ancestors of the Hindu-Arabic system was already common. Because Āryabhata was writing a poem, however, that form of number was not convenient for fitting into the meter. As a result, he introduced a system of words in the first section of the poem to correspond to numbers. His numerical vocabulary therefore is not as wide as the range of numbers already in use in Indian computation but sufficed for the writing of his poem.

After the opening invocation in the section devoted to mathematics (*ganita*, in Sanskrit), Āryabhata lists in the second stanza the names for classes of numbers. The third stanza is on the subject of geometry; it points out that the product of two equal numbers was called by the same name as both the square of a figure and its area. He points that the same consideration applies in three dimensions to the cube and the product of three equal quantities.

The following two stanzas (four and five) present methods for calculating the square root and the cube root of a number. From the form of the instruction, it is clear that Āryabhata recognized that the square root and cube root were not generally going to terminate in a finite number of places. The geometrical points made in the previous stanzas could have come out of the Greek tradition of theoretical mathematics, but these techniques for calculation could be part of a native Indian tradition. In general, the issue of how far Āryabhata was indebted to Greek sources is hard to resolve, and the discussion often revolves about political issues as well as mathematical or scholarly ones.

The next four stanzas all take up the area and volume of various two-dimensional and three-dimensional figures. The formulae for the area of a triangle and a circle are correct, as is that for the volume of a pyramid, but the formula for the volume of a sphere is incorrect. If Āryabhata were drawing on the Greek geometrical tradition throughout, it is hard to see how he could have made this error, which seems to be based on an analogy with the area of a circle. In addition, he gives the area for a trapezoid and plane figures more generally. The more general formulae may be suggested by the kind of averaging involved in handling the trapezoid. It is worth remembering that all the so-called formulae are really expressed in the language of poetry rather than written out as equations.

The tenth stanza includes one of the most interesting features of *The Aryabhatiya*, an approximation for π (the ratio of the circumference of a circle and its diameter). The specific numerical values given by Āryabhata lead to a decimal approximation good to four places, and there is plenty of discussion about how this compares with the best possible values that the Greeks obtained. In principle, for example, the method of exhaustion of Archimedes (c. 287-212 B.C.E.) could produce approximations to an arbitrary degree of exactness. What is clear is that Āryabhata recognized that his value remained only an approximation.

The following two stanzas deal with the calculation of quantities related to the trignometric sine function. The centrality of the sine function in the mathematics of Āryabhata came from its appearance in the tradition of calculating orbits of planets that goes back to Greek explanations of observations, records of which survive from Babylonian times. The mathematical astronomy of Ptolemy (c. 100-178 C.E.) was probably brought to India in the course of the centuries before Āryabhata's work, as there were Greek and Egyptian traders in India and Indian traders in the centers of culture of Europe and the Mediterranean. However, there was also a native tradition of astronomy going back many centuries, although it is not clear how mathematical a form it had taken. Āryabhata's interests seemed to merge those of Greek mathematics and Hindu astronomy.

Greek mathematics took issues of construction as seriously as proofs, and Āryabhata includes a stanza on the construction of various figures. The next three stanzas take up matters relating to the construction of sundials, not surprising in view of the section following that on mathematics being devoted to time measurement. The seventeenth stanza then turns to the Pythagorean theorem about the relationship between the sides of a right triangle. Historians have found references to the theorem in many mathematical cultures other than the Greek, and it could easily have been part of an Indian tradition as well. Because there is no reference to a proof, it is hard to see any connection with the version given in Euclid (c. 330-c. 270 B.C.E.).

After a stanza on the lengths of chords in a circle, Āryabhata explains how to calculate the sum of the terms in an arithmetic series (where the difference between successive terms is constant). He also provides the way to calculate the number of terms involved if one has the sum, and this presupposes a knowledge of the quadratic formula, the way to derive the roots of a quadratic (second-degree) equation. The quadratic formula had been used by Babylonian mathematicians much earlier, and it is hard to tell how it fit into the rest of Āryabhata's algebraic knowledge. The formulae he then proceeds to give for the sum of the squares and cubes of arithmetic series also come without derivation.

A couple of stanzas are devoted to getting the values of individual terms from various ways in which they can be combined. After a stanza (number 25) on interest (in which compounding takes place), Āryabhata introduces the rule of three, the standard method of dealing with proportions for algebraists for millennia. In subsequent stanzas, he provides a guide to calculating common denominators for fractions and resolving problems by using the inverse of the operations in which the problems are posed. The range of subjects continues to widen, as he gives a way of getting a sum from differences, of calculating the value of an object with an agreed monetary standard, and of calculating the distances between planets.

The height of Āryabhata's mathematical originality comes in the final two stanzas, in which he presents the general techinique for solving Diophantine equations of the first degree. Diophantus (fl. c. 250 C.E.) had presented solutions of number of special cases, but Āryabhata clearly developed more general forms. The additional detail given to the method (by taking an extra stanza) and its position at the end of the section suggests that Āryabhata recognized the extent of what he had accomplished.

The third section of *The Aryabhatiya* looks at issues relating to time but largely through the medium of astronomy. The final section takes up the celestial sphere, and both sections had an influence on subsequent Indian astronomy as well as being brought subsequently to the West. The work as a whole ends with a couple of stanzas celebrating his own accomplishments and criticizing those who disparage the field of astronomy. It is safe to say that this was in response to critics, but it is not clear what the basis was for their objections.

After Āryabhata's death, subsequent Indian mathematicians, especially Brahmagupta (c. 598-c. 660), disparaged his accomplishments for not having been presented systematically. On the other hand, his work remained the starting point for Indian mathematics and astronomy for many years. Commentaries were published on *The Aryabhatiya* for more than a thousand years, and it has also drawn the attention of modern scholars. As historians of mathematics look for sources outside the Greek tradition, Āryabhata's work is both attractive for its style and worthy of investigation.

Significance

Āryabhata stands at the head of the Indian tradition of mathematics. There was clearly a stream of predecessors on which he drew, but their anonymity makes their importance a trifle elusive. The work of Āryabhata provided a spur for a thousand years of mathematical research. The form in which he presented his work also helped make sure that it could be taken as a classic contribution to Indian scholarship, not merely a mathematical thesis.

When India launched its first satellite in 1976, it was the 1500th anniversary of Āryabhata's birth, and the satellite was named after him. At present, there is an Āryabhata Forum centered in the United Kingdom working on issues in the history of mathematics. The legacy of Āryabhata continues to serve as a reminder that not all mathematics can be assumed to have come through the Greek pipeline.

Thomas Drucker

Further Reading

Menon, K. N. *Aryabhata: Astronomer, Mathematician.* New Delhi, India: Ministry of Information, 1977. Part of the spate of literature produced by the 1500th anniversary of Āryabhata's birth, this work is typical in that it claims all sorts of achievements for Āryabhata at the expense of both other mathematical cultures (like the Greek) and other Indian mathematicians.

Proceedings of the International Seminar and Colloquium on Fifteen Hundred Years of Aryabhateeyam. Kochi, India: Kerala Sastra Sahitya Parishad, 2002. This was sparked by the 1500th anniversary of Āryabhata's book rather than his birth. The collection of articles covers a good deal not connected with Āryabhata, but it does provide evidence for continued disagreements over some of the basic facts of his life.

Velukutty, K. K. *Heritages to and from Aryabhatta.* Elipara, India: Sahithi, 1997. Another assault on the standard accounts of the details of Āryabhata's life. It follows up some other Indian mathematical traditions with attention to details of practice.

WALTER BAADE

German astronomer

Baade was one of the most influential astronomers of the twentieth century. He made major discoveries that led to greater scientific understanding of stellar and galactic evolution, was instrumental in the discovery of the supernova phenomenon, and introduced the concept of stellar populations.

Born: March 24, 1893; Schröttinghausen, Westphalia, Germany
Died: June 25, 1960; Bad Salzuflen, Westphalia, Germany
Also known as: Wilhelm Heinrich Walter Baade (full name)
Area of achievement: Astronomy

Early Life

Walter Baade (BAH-duh), christened Wilhelm Heinrich Walter Baade, was the oldest son of Konrad Baade and Charlotte Baade. Walter suffered a hip disorder from birth that plagued him his entire life. Konrad was a schoolmaster, and had hopes of his son studying theology and entering the clergy. From 1903 to 1912, the young Baade studied at Friedrichs Gymnasium (a German secondary school), where he showed a strong aptitude for science. He then spent a year at the University of Münster before transferring to the University of Göttingen in 1913.

Baade continued studies at Göttingen until 1919, when he received his Ph.D., with his doctoral thesis looking at the eclipsing binary star Beta Lyrae. His hip injury exempted Baade from military service in World War I, though he was pressed into service from 1916 to 1918 performing aerodynamics weapons research. The Göttingen Observatory was poorly equipped, and Baade did not get much observing experience. After receiving his Ph.D., Baade took a position at the Hamburg Observatory in Bergedorf, operated by the University of Hamburg, where he became an expert in observational astronomy. Baade married Johanna "Muschi" Bohlmann, a computational assistant at the University of Hamburg, in 1929. Asteroid 966 Muschi was named for her.

Life's Work

Baade became a well-known astronomer because of his work at Bergedorf, where he discovered Comet Baade (C/1922 U1) and the asteroid 944 Hidalgo. For many years 944 Hidalgo was the most distant asteroid discovered; it had an orbit that ranged nearly as far as Saturn. Baade secured a Rockefeller Fellowship to travel to Harvard, Yerkes Observatory in Wisconsin, and Mount Wilson Observatory in Southern California for a year beginning in 1926. Then, in 1929, Baade traveled together with famed optician Bernhard Voldemar Schmidt to the Philippines to observe a solar eclipse. In 1931, Walter Adams of Mount Wilson Observatory, greatly impressed with Baade from his earlier work there, offered him a permanent staff position. Baade quickly accepted, and he and his wife moved to California, where he would remain for most of his life.

At Mount Wilson, Baade began work with Fritz Zwicky studying a certain type of nova that he had referred to as a *hauptnova* (chief nova) in a lecture that Baade had given at Hamburg a few years earlier. The two astronomers realized that novae of this type were far too luminous to be anything other than an exploding star. Together, they worked out the physics of stellar explosions and the phenomenon that is now called a "supernova," a term Baade and Zwicky coined to replace Baade's earlier *hauptnova* designation.

By the late 1930's, the Nazis had secured much power in Germany. Baade worked very hard to help a Jewish friend and colleague, Rudolph Minkowski from the University of Hamburg, escape from Germany and the Nazis. Disgusted with what was happening in Germany, Baade reportedly initiated an application for U.S. citizenship, but the paperwork was lost. Scornful of the bureaucracy that had lost his paperwork, Baade never reapplied. Thus, when the United States entered World War II, Baade, still a German citizen living in California, was declared an enemy alien. However, his colleagues at Mount Wilson managed to keep him from being sent to an internment camp by ensuring the U.S. government that he would remain in the immediate vicinity of Mount Wilson for the duration of the war.

With most of the observatory's scientists performing research elsewhere in support of the war effort, Baade had the place to himself. Adding to this ideal situation for Baade was that the greater Los Angeles area, which Mount Wilson overlooks, was under blackout conditions early in the war because of fears of a possible Japanese invasion along the West Coast. The night skies at Mount Wilson were the darkest they had been since the observatory was built. Under such dark skies, Baade began studies of the Andromeda galaxy. He found that there were two distinct types of stars in the galaxy. One type, which he called Population I, was found in the spiral arms of the galaxy and was bluer than the other type. The second type, which he called Population II, was found in the central regions of the galaxy, in its globular clusters, and in its elliptical satellite galaxies. Population II stars were much redder and dimmer on average than the Population I stars. More research showed that Population I stars are far richer in metals than are Population II stars. (Astronomers call anything other than hydrogen and helium a "metal.") Theoretical studies later showed that this implies that Population I stars formed more recently and Population II stars formed earlier in the history of the galaxy. Researchers can make this implication because metals are produced in the fusion reactions in the cores of stars and are seeded into space when stars die, which allows them to become constituents of the next generation of stars.

Baade next turned his attention to Cepheid variable stars. Cepheid variables are pulsating stars whose pulsation period is related to their luminosity. The brighter Cepheids pulsate slower than the dimmer ones. This relationship can be used to determine distances to star clusters and nearby galaxies. Baade's studies showed that Cepheids fall into two categories, based upon their stellar population. By placing them into their separate types, Baade was able to improve the period-luminosity relationship and thus also provide for improved distance determinations. He found that nearly all distances to galaxies had been underestimated by a factor of at least two. The universe was, therefore, far larger, and far older, than had been thought. This was great news for geologists, for they had been coming up with ages for the earth that were billions of years older than cosmologists were computing for the universe. After Baade's correction, the two ages were closer in agreement. This also provided an age for the universe more consistent with that needed for George Gamow's big bang cosmology.

Baade finally retired in 1958. However, retirement did not stop him from working. In 1959, Baade taught a graduate class at Harvard on stellar and galactic evolution and how those topics can be better understood using stellar populations. Later, Cecilia Payne-Gaposchkin edited transcripts of his lectures and they were published as *Evolution of Star and Galaxies* (1963). Baade then flew to Australia, where he traveled and repeated his Harvard lectures.

Finally, in late 1959, Baade returned to his beloved boyhood home of the Westphalia region of Germany. He had plans to write some much delayed papers and to teach at the University of Göttingen. He also was considering doing more work at Hamburg's Bergedorf Observatory. However, his congenital hip disorder had become far worse, and he was in constant pain. By early 1960, the condition had progressed to where he was beginning to suffer paralysis of his legs. He underwent major surgery to correct some of the problems. He was instructed to remain in bed for five months after the surgery, a "confinement," as he called it, that was very difficult for him. Finally, he was allowed to sit up in a wheelchair, but he died days later, on June 25, in Bad Salzuflen. It has been speculated that perhaps a blood clot may have formed as a complication of his surgery and the five-month convalescence. Sitting up may have dislodged the clot.

Significance

Baade published less than many other astronomers. However, nearly every paper that he did publish had a major impact on astronomy and astrophysics, helping to shape twentieth century astronomy. Often regarded by some as "just" an observational astronomer, Baade's grasp of astrophysics was extraordinary.

Baade's discovery of Population I and Population II stars was one of the most important discoveries in stellar astronomy. It set the stage for astrophysicists to understand stellar and galactic evolution. Modern astronomical thought no longer groups stars into only two distinct populations. Rather, stars are now thought to lie along a continuum of metallicities, that is, the constitution of a star varies in its amount of matter beyond the chemical elements helium and hydrogen. Nonetheless, the terms "Population I" and "Population II" are still used by astronomers, and the metallicity of a star is considered an important measurement.

Baade's, and Zwicky's, research on supernovae was a major step forward in understanding the evolution of high-mass stars, and his work with distance scales also was instrumental in supporting the big bang cosmology and putting the steady state model to rest.

Raymond D. Benge, Jr.

Further Reading

Baade, Walter. *Evolution of Stars and Galaxies*. Cambridge, Mass.: Harvard University Press, 1963. A transcript of a series of lectures given by Baade, edited by Cecilia Payne-Gaposchkin, that clearly explain Baade's work in stellar evolution and galactic astronomy.

Osterbrock, Donald E. *Walter Baade: A Life in Astrophysics*. Princeton, N.J.: Princeton University Press, 2001. A well-researched, recommended biography of Baade by a historian of astronomy. Includes a bibliography.

_______. "Walter Baade, Master Observer." *Mercury* 31, no. 4 (July/August, 2002): 32-41. A brief biography of Baade, focusing mostly on the highlights of his life.

JEAN-SYLVAIN BAILLY

French astronomer and politician

Bailly, a renowned astronomer and historian of science, was elected president of the Third Estate in 1789. In the same year, he became the first mayor of Paris under the New Republic. He soon fell out of favor, however, and was later guillotined by the Revolutionary Tribunal of Paris.

Born: September 15, 1736; Paris, France
Died: November 12, 1793; Paris, France
Areas of achievement: Science and technology, astronomy, historiography, government and politics

Early Life

Jean-Sylvain Bailly (zhahn-seel-va bah-yee) was born into a family of painters to the royal court in Versailles. His father was court painter and custodian of the royal art collection at Versailles. It was originally understood that

Jean-Sylvain would apprentice under his father and inherit his post, but he showed little interest in art. Instead, he showed a strong aptitude for mathematics, and through a special arrangement he was tutored in calculus by Monsieur de Moncarville, whose son was in turn tutored in art by Jean-Sylvain's father.

Bailly also had literary aspirations and as a teen penned two plays. The playwright Lanoue, after being shown the plays, told Bailly they were so poor he should burn them. From this point forward, Bailly focused most of his attention on the sciences and was soon drawn especially to astronomy by Abbé Nicolas Louis de Lacaille.

Life's Work

Bailly made his first mark in astronomy by calculating the orbit of Halley's comet when it appeared in 1759, a clear sign of his fascination for the mathematical side of astronomy. He calculated the orbits of several lesser comets as well. As a result of these and other astronomical observations and calculations, he was elected to the French Academy of Sciences in 1763.

Bailly continued his astronomical work, studying the orbits of the moons of Jupiter. It was well known that, despite careful observation, there were inconsistencies in their orbital paths. A number of astronomers had tried mathematical solutions. Bailly's solution was elegant, considering the knowledge of the time (Laplace later found a better solution). He published two books dealing with the moons of Jupiter: *Essai sur la théorie des des satellites de Jupiter* (1766; essay on the theory of the satellites of Jupiter) and *Mémoires sur les inégalités de la lumière des satelites de Jupiter* (1771; memoirs on the uneven illumination of the satellites of Jupiter).

Bailly's interests also strayed into questions concerning the ultimate fate of the Earth and the other planets. He postulated that each planet has an inner heat that is at some stage of cooling. Jupiter he considered too hot for life, and the Moon was too cold. Because the cooling was a continuing process, he believed that all the planets would eventually cool to the point where all matter would cease movement.

As his renown grew, Bailly was admitted to the French Academy and the Academy of Humanities. He developed a long-term relationship with Voltaire, later publishing many of the letters they exchanged. He also met Benjamin Franklin, a fellow scientist and politician.

During the 1770's and 1780's, Bailly published a number of books, including his most famous work, a three-volume history of modern astronomy, *Histoire de l'astronomie moderne* (1779-1782). This work followed his earlier *Histoire de l'astronomie ancienne* (1775; history of ancient astronomy). In a book on Indian and Oriental astronomy, he showed that the ancient Vedic astronomical tables not only were very accurate but also were probably of ancient origin. He was also fascinated with Plato's writings on Atlantis in the *Critias* (360-347 B.C.E.; English translation 1793), and he wrote a book resurrecting Plato's theories about the origin of Atlantis, *Lettres sur l'Atalantide de Platon et sur l'ancienne histoire de l'Asie* (1779; letters on Plato's Atlantis and the ancient history of Asia). Bailly also speculated that the ancient Eden of the Bible was polar and claimed evidence from ancient Egyptian and Syrian manuscripts as support.

Jean-Sylvain Bailly. (Library of Congress)

In 1789, the French Revolution exploded, bringing an abrupt change to Bailly's life. Because of his fame as a scientist and his high level of integrity, he was catapulted into politics. On May 5, 1789, Bailly was elected president of the Third Estate, the assembly (somewhat like the British parliament) that represented the larger part of the French population. As a member of the bourgeoisie, the upper-class citizenry that often served on the Third Estate, he served as a political link between the working classes and the First and Second Estates (the clergy and the nobility, respectively). On July 15, Bailly was again elevated, this time to mayor of Paris, the city's first mayor of the New Republic.

With no experience as a politician, Bailly was now thrust into one of the most difficult political situations imaginable. Because of the new humanistic ideas of the Enlightenment, the expansion of the middle class leading to clashes with the aristocracy, and the serious governmental financial crisis, Paris was thrown into anarchy. The only advantage Bailly had entering office was that he was on relatively good terms with most of the political players from the French monarchy on down.

Conditions in Paris were unmanageable. The peasants and middle class had overthrown the city's police powers by storming the Bastille, Paris's notorious prison-fortress. With no police protection, many merchants' shops were looted. Commerce came to a standstill. To make matters worse, the grain harvest had been very poor, and the city was short on bread. With the shortage, bakers raised bread prices to exorbitant heights, reflecting their increased costs but further enraging the populace.

At first, Bailly felt honored to serve as mayor, but reality soon set in. The first order of business was to secure the food supply, the most inflammatory source of unrest. To do this, Bailly and his administration enacted price caps on bread so the working classes could afford to eat. Then he began to troll far and wide for grain sources, even to the point of buying from other countries. At any given time, there was often only about a day's supply of grain in the city. Bailly managed to keep the precarious nature of the situation from the people, hoping it would prevent further rioting.

The other pressing need was to restore order through reforming the municipal government of Paris. The previous administration was so bureaucratically complex that no progress could be made, so a completely new constitution had to be enacted. Even this process was fraught with difficulty, as each of the districts of the city wanted to retain autonomy. Bailly was a strong proponent of a centralized government with the mayor having broad powers, whereas the elected assembly he had to work with wanted a much-less-centralized power structure. Eventually, Bailly had to compromise, with the result that his office as mayor had very little power: The right to enact laws was retained by each of the city's departments.

In spite of Bailly's initial success, the instability of Paris grew worse. Establishment of a national guard was helpful, but when "royal troops" were sent to Paris with orders to kill civilians and then fall back, Bailly saw no choice but to respond more forcefully to the unrest. On July 17, 1791, Bailly imposed martial law. Rioting broke out everywhere, and it was on this fateful day that General Lafayette ordered his militia to fire on an unruly mob at the Champs de Mars. Twenty-four civilians were killed in the attack. Lafayette went into exile two days later, and Bailly's reputation was irreparably damaged.

On September 19, Bailly was forced to resign as mayor. Refusing to exile himself, he left Paris and moved to Nantes, where he spent the next couple of years writing his memoirs. Late in 1793, he was arrested while visiting Laplace in Melun. He was tried on November 10 by the Revolutionary Tribunal. Two days later, on November 12, 1793, Bailly was led to the Champs de Mars so he could be executed on the spot where his crime was committed. He was marched before a jeering, rock-throwing mob for several hours and was later led to a place near the Seine, where he was guillotined.

Significance

Jean-Sylvain Bailly's place in history has been nearly forgotten. His accomplishments in astronomy, although notable at the time, have lost much of their importance. His accomplishments in the history of astronomy have been more lasting, however, and his book on the subject remains a classic. His astronomical achievements are permanently commemorated by Bailly Crater, the largest crater on the near side of the moon.

Bailly's greater significance was certainly the role he played in the French Revolution. Although a man of science, he was also a man of deep moral convictions. This trait was recognized by those who helped propel him into politics. His inexperience was both an advantage and a disadvantage. Not having ever served in office before, he was unsullied by the political machinations so common in politics, thus gaining the trust of the populace. On the other hand, his lack of experience made him an easy target for those who were more experienced at the game. In some ways, it is surprising that he survived in his role as mayor as long as he did.

Bailly's lasting legacy lies in the fact that he stood by the people of Paris in one of their darkest hours. He was a voice of reason between the extremes of anarchy and monarchy. He helped shape the constitution that lay at the foundation of the New Republic. He also paid the ultimate price for his role. He could have gone into exile to Great Britain or elsewhere, but he chose to remain in France, where, at the age of fifty-seven, he was executed in the city he loved.

Bryan D. Ness

Further Reading

Brucker, Gene A. *Jean-Sylvain Bailly: Revolutionary Mayor of Paris*. Urbana: University of Illinois Press, 1950. A short overview of Bailly's life during his political career from his election to the presidency of the Third Estate to his resignation as mayor of Paris.

Brush, Stephen G. *A History of Modern Planetary Physics*. 2 vols. New York: Cambridge University Press, 1996. Contains a few brief sections discussing some of Bailly's astronomical research and theories.

Pulet, Ann L. *Jean-Antoine Houdon: Sculptor of the Enlightenment*. Chicago: University of Chicago Press, 2005. Contains pictures and discusses the bust sculpted in honor of Bailly's election as mayor of Paris.

BENJAMIN BANNEKER

American mathematician and astronomer

Banneker's calculations provided the essential data for almanacs published from 1792 through 1797. A free black in a slave state, Banneker overcame obstacles of rural isolation, little formal education, and racial prejudice to establish himself as a respected scientist, earn a place on the crew that surveyed the District of Columbia, and become a symbol of racial equality in the abolitionist movement. He is also known for his letter to Thomas Jefferson calling on him to end slavery in America.

Born: November 9, 1731; Baltimore County, Maryland
Died: October 9, 1806; Baltimore, Maryland
Areas of achievement: Mathematics, astronomy, science and technology

Early Life

Benjamin Banneker's American antecedents came in bonds to colonial Maryland. His grandmother, Molly Welsh, was a convict transported from England to Maryland in about 1683. After completing a period of servitude, she became a free landowner in the western part of Baltimore County near the Patapsco River. In 1692, Molly bought two Africans and in a few years restored freedom to both. One of the men, named Bannka, claimed to be the kidnapped son of an African king. In defiance of laws that forbade miscegenation, Molly married the prince and took Banneky as her surname.

The Bannekys had four daughters. The oldest, Mary, born in about 1700, married an African who had been given freedom as a baptismal gift. He had chosen Robert as his Christian name and, when married, took Banneky as his surname. The name's spelling varied until the mid-eighteenth century, when it settled at Banneker. Three of the four children born to Robert and Mary grew to maturity. The oldest, and the only son, was Benjamin, born in 1731.

In about 1729, Robert bought 25 acres of land close to Molly's farm. On March 10, 1737, when Benjamin was five years old, Robert purchased 100 acres from the nearby Stout plantation. The title was in Robert's and Benjamin's names to ensure that the family could protect its freedom should Robert die suddenly. Maryland laws were not sympathetic to free blacks and authorized reenslavement of those who did not own property.

Banneker's education was rudimentary. His grandmother taught him to read from the Bible. For a few months, he attended a country school where the schoolteacher, probably a Quaker, taught black and white children. Benjamin learned to write a very clear, even beautiful, script and mastered the fundamentals of mathematics through basic algebra. At some point, he also learned to play the flute and the violin. Though meager, this education powerfully shaped the course of Banneker's life. He purchased his own Bible in 1763, read it diligently, and sprinkled his writings with scriptural quotations. He never formally joined a Christian denomination, but he often attended Quaker, and sometimes Methodist, services. His reading interests went beyond the Bible to literature in general. He painstakingly compiled a small library, composed essays in his own commonplace book, and wrote poetry.

Mathematics, though, was the subject that most stimulated his intellectual curiosity. He had unusual abilities with numbers. As a young man, he became locally famous for being able to solve fairly complex computations in his head. He had a special fondness for mathematical puzzles and liked to trade tricky problems with his neighbors.

It was probably during such an exchange with a neighbor that Banneker first saw and then borrowed a watch. The timepiece fascinated him, and he dismantled it to observe its moving parts. Using the watch as a model, Banneker produced a clock made entirely of hand-carved hardwoods. The clock kept accurate time, struck the hours, and was the wonder of the Patapsco valley.

Banneker completed the clock in 1753, when he was twenty-two. His father died six years later, leaving Benjamin the sole owner of the Stout acreage. The rest of the property was divided among Benjamin and his two married sisters. Banneker lived with his mother until she died in 1775. He never married and lived the rest of his life on his well-kept, productive farm. He might have died in obscurity had not the Ellicott brothers bought land adjoining the Banneker farm.

Life's Work

Joseph, Andrew, and John Ellicott brought their large families and the families of several workers to the Patapsco valley in 1771, when Benjamin Banneker was about forty years old. Before they were fully settled, the Ellicotts and their workers bought food from the existing farms. Andrew Ellicott's young son George developed a special friendship with Banneker. At age fifteen in 1775, George was recognized as a mathematical prodigy, an accomplished surveyor, and a gifted astronomer. With George's encouragement and assistance, Banneker rapidly mastered advanced mathematics and became fascinated with astronomy. In the fall of 1788, George loaned Banneker books on mathematics and astronomy, a telescope, a set of drafting instruments, a lamp, and a large, oval drop-leaf table.

Banneker soon spent clear nights in open fields observing the heavens. When recording his observations in cold weather he dressed heavily and wrapped himself in blankets. At dawn he returned to his cabin and slept for a few hours. He spent most of the rest of the day at the oval table studying the borrowed books and plotting the movements of the stars. The calculation of a star's location for a particular date could involve as many as ten different algebraic and logarithmic operations.

As he gained a sure grasp of astronomy, Banneker began the ambitious project of calculating an ephemeris—a table showing the positions of the Earth, Moon, planets, and stars throughout the year. The ephemeris was the basis for projecting eclipses and predicting weather conditions. It was, therefore, the major component of an almanac, a compilation of astronomical and weather-related data for a given year. He was encouraged in this project by his mentor George and by George's cousin, Major Andrew Ellicott. Major Ellicott had prepared ephemerides for publication from 1781 through 1786, but the demands of his work as surveyor forced him to abandon the time-consuming calculations. In the summer of 1790, Banneker submitted completed ephemerides for 1791 to three publishers, but none of the editors bought his work. Although discouraged, Banneker began an ephemeris for 1792.

In January of 1791, President George Washington instructed Secretary of State Thomas Jefferson to have the District of Columbia surveyed. On February 2, 1791, Jefferson named Major Ellicott the chief surveyor. Ellicott was to find the true meridian and longitude of the future capital and to prepare a topographical map of the ten-square-mile tract of land. Because he was shorthanded, Ellicott turned to Banneker for help. Banneker had no practical experience as a surveyor, but he had mastered the mathematics involved and knew how to work with most of the astronomical instruments.

On February 7, 1791, Banneker, at age fifty-nine, made his first trip outside Baltimore County. During his three-month stay at the site of the future capital, he gained valuable experience as assistant to Ellicott. He learned to use the astronomical clock and other instruments new to him. He also kept a resolution to abstain from drinking wine and hard liquor while working with the surveying crew.

Upon his return home, he finished the ephemeris for 1792. Meanwhile, Joseph and George Ellicott had interested members of the Society of Friends and the antislavery societies in Baltimore and Philadelphia in Banneker. Through their assistance, Banneker's 1792 ephemeris was published by the Baltimore firm of Goddard and Angell. The almanac, *Benjamin Banneker's Pennsylvania, Delaware, Maryland and Virginia Almanack and Ephemeris for the Year of Our Lord, 1792* was sold in Baltimore, Philadelphia, and Alexandria beginning in December, 1791. The editors of the almanac included a note in the work's front matter, stating that the almanac was "an extraordinary Effort of Genius—a COMPLETE and ACCURATE EPHEMERIS . . . calculated by a sable Descendant of Africa. . . ." The editors argued that the almanac was proof that skin color had no relationship to mental or intellectual capacity, that all people were alike, and that slavery should be ended. The first four thousand copies quickly sold out, as did a second printing by Goddard and Angell and a condensed edition printed by William Young in Philadelphia.

Banneker prepared an ephemeris that was published in an almanac each year through 1797. The almanacs were extremely popular and sold well. At least twenty-eight editions of these almanacs appeared in those six years. Starting in 1794, Banneker computed tide tables for Chesapeake Bay, a feature that competing almanacs did not contain.

The elderly astronomer was of average height and had a full head of thick, white hair. Though portly, his posture, gentlemanly behavior, and staff gave him a dignified air. Banneker continued to calculate ephemerides through the 1802, the year he turned seventy-one. His capacity for work had diminished, and he was unable to complete the rigorous computations. He died quietly in his home four years later on October 9, 1806.

Significance

Benjamin Banneker's abilities as a mathematician and astronomer made him famous in his lifetime. There was, however, much more to his fame than his scientific accomplishments themselves. Banneker was the son and grandson of Africans. He was a free black in a predominantly white society that almost universally regarded black people as being mentally inferior to whites. Banneker's best-known activity—in his lifetime and since—was his correspondence with Thomas Jefferson.

On August 19, 1791, Banneker sent the author of the Declaration of Independence a handwritten copy of his ephemeris and a long letter. He introduced himself as a black man and then eloquently pleaded with Jefferson to use his influence to end the slavery that still kept some of the children of humankind's one Father from enjoying their "inalienable rights." Jefferson responded by expressing the hope that people would soon recognize that circumstances, not natural endowments, kept blacks in a condition that suggested inferiority, but he made no pledge to do anything more than to send Banneker's almanac to abolitionists in France.

Those two letters were printed in the 1793 almanac and reprinted frequently in abolitionist literature in the nineteenth century. Benjamin Banneker had become a symbol of racial equality because he was an example of black achievement. His name has been invoked over the years in black educational efforts, such as Benjamin Banneker College of Prairie View A&M University at Prairie View, Texas.

Paul E. Kuhl

Further Reading

Allen, Will W., comp. *Banneker: The Afro-American Astronomer*. Washington, D.C.: Black Heritage Library, 1921. Reprint. Freeport, N.Y.: Books for Libraries Press, 1971. This work is based largely on primary sources. However, the book contains a paper by Daniel Murray that advances the oft-repeated, undocumented claim that Banneker worked with Major Pierre Charles L'Enfant and had a copy of the city's plans for use after L'Enfant left the capital site in a rage.

Allen, William G. *Wheatley, Banneker, and Horton*. Boston: D. Laing, 1849. Reprint. Freeport, N.Y.: Books for Libraries Press, 1970. A condensation of an earlier work on records held by the Ellicott family.

Armistead, Wilson. *A Tribute for the Negro: Being a Vindication of the Moral, Intellectual, and Religious Capabilities of the Colored Portion of Mankind, with Particular Reference to the African Race*. 1848. Reprint. Miami, Fla.: Mnemosyne, 1969. A collection of abolitionist literature. The material on Banneker consists of a brief biographical sketch and a reprint of the Jefferson correspondence.

Baker, Henry E. "Benjamin Banneker, the Negro Mathematician and Astronomer." *Journal of Negro History* 3 (April, 1918): 99-118. A fine sketch based upon a rare work prepared by George Ellicott's daughter Martha Ellicott Tyson.

Bedini, Silvio A. *The Life of Benjamin Banneker: The First African-American Man of Science*. 2d rev. ed. Baltimore: Maryland Historical Society, 1999. An excellent study of Banneker, based on careful review of secondary materials, previously unused material from private archives, and Banneker's commonplace book and journal. Originally published in 1971, this expanded edition includes new photographs and information on Banneker's African roots.

Cerami, Charles. *Benjamin Bannecker: Surveyor, Astronomer, Publisher, Patriot*. New York: J. Wiley, 2002. A clearly and simply written biography that examines Banneker's life and the breadth of his career.

Miller, John Chester. *The Wolf by the Ears: Thomas Jefferson and Slavery*. New York: Free Press, 1977. Miller treats the correspondence between Banneker and Jefferson in the context of Jefferson's life and slavery in the United States.

AL-BATTANI

Arab astronomer and mathematician

Al-Battani examined and corrected, through application of trigonometry, astronomical theories first put forward by the second century Alexandrian Ptolemy.

Born: 858; near Haran, Mesopotamia (now in Turkey)
Died: 929; Qasr al-Jiss, region of Samarra (now in Iraq)
Also known as: Abū 'Abd Allah Muhammad ibn Jabir ibn Sinan al-Battani al-Harranī al-Sabi' (full name); Albatenius; Albategnius; Albategni (Latin names)
Areas of achievement: Astronomy, mathematics, science and technology

Early Life

Born near Haran, the young al-Battani (ahl-bah-TAH-nee) moved with his and several other families to Rakka on the Euphrates River midway on the caravan route between Aleppo and the Upper Mesopotamian city of Mosul. This migration may be explained in part by the *nisba*, or nickname, retained by the future Islamic astronomer. "Al-Sabi'" may refer to his family's earlier adherence to the so-called Sabian sect, which was reputed to follow a mixture of Christian and Islamic principles. Whatever the family's original religious orientation, al-Battani's later fame was won under the banner of Islam, the faith he ultimately followed. After his move to Rakka as a youth, al-Battani spent the remainder of his life in the same geographical and cultural environment.

No specific information is available on al-Battani's formal education. It is not known, for example, whether his original training was obtained in a fully secular "scientific" or in a religious setting. It was as a youth in Rakka, however, that al-Battani decided to devote himself to careful study of ancient texts, especially those of Ptolemy, which provided him with the knowledge needed to carry out the series of astronomical observations that would make him famous, not only in the Islamic world but in the medieval European West as well.

Life's Work

Al- Battani, known to the West as Albatenius, contributed greatly to advances in the field of trigonometry. To carry out key calculations, he relied on algebraic rather than geometric methods. Like his somewhat lesser known later follower Abul Wefa (940-998), al-Battani focused much of his attention on the theories of the second century Alexandrian astronomer Ptolemy. Several Islamic scholars before him had been intrigued by Ptolemy's approach to the phenomenon of the oscillatory motion of the equinoxes. Al-Battani's contemporary Thabit ibn Qurrah (c. 836-901) tried to account for this by supplementing Ptolemy's theory, merely adding a ninth sphere to the Greek scientist's assumption of eight spheres; al-Battani, however, remained doubtful. He was convinced that trigonometry should be developed more effectively for the purpose of achieving greater precision in already known methods of making these and other astronomical calculations. This goal led him to explore and expand the relevance of sines. His use of the Indian sines, or half chords, enabled him to criticize Ptolemy's conclusions in several areas.

For example, Ptolemy had insisted that the solar apogee was a fully immobile phenomenon. Al-Battani, however, was able to observe that in the seven centuries since Ptolemy's time, there had been a notable increase in the Sun's apogee. His further observations suggested that the apogee was affected by the precession of the equinoxes. To explore this theory required a substantial revision of methods of proposing equations to represent the passage of time in accurate astronomical terms; room had to be made for accommodating slow secular variations. As part of this process, al-Battani set out to correct Ptolemy's theory of the precession of the equinoxes.

The phenomenon of eclipses was also a field incompletely pioneered by Ptolemy. Interest in this subject motivated al- Battani to make a variety of studies that aided subsequent astronomers in their calculations to determine the time of the visibility of the new moon. His treatment of the phenomena of lunar and solar eclipses provided the basic information that would be used by European astronomers as late as the eighteenth century. Most notably, Richard Dunthorne used al-Battani's work in his 1749 study of the apparent acceleration of the motion of the Moon. In addition, mention of solutions al-Battani proposed for the field of spherical trigonometry appears in many earlier European works, including those of Regiomontanus (1436-1476).

In a somewhat more practical vein easily appreciated by the layperson, al-Battani's observations allowed him to determine the length of the tropic year and, significantly, the precise duration of the four seasons of the year.

One of the most original areas of al-Battani's work involved the use of horizontal and vertical sundials.

Through their use, he was able to denote the characteristics of a so-called "horizontal shadow" (*umbra extensa*). These he used to reveal cotangents, for which he prepared the first-known systematic tables. Similarly, his study of "vertical shadows" (*umbra versa*) provided pioneer data for calculating tangents.

Most of al-Battani's important findings in the field of astronomy were contained in his major work, *Kitab al-zij* (c. 900-901, book on astronomy; best known as *Zij*, *De motu stellarum*, or *De scientia stellarum*). As the Latin titles suggest, this magnum opus was first circulated widely among scholars of the early period of the European Renaissance. The work was translated originally by Robertus Retinensis in Spain in the twelfth century. In the thirteenth century, King Alfonso X of Spain had a direct Arabic-to-Spanish translation prepared. *De scientia stellarum* later gained added attention from modern scholars such as the Italian C. A. Nallino, who edited and translated the Latin text, providing essential commentaries that enhance contemporary understanding of the Islamic scholar's original contributions.

Unfortunately, modern scholars' familiarity with other important writings by al-Battani is limited to what can be gleaned from references to them in other Islamic authors' works. A "Book of the Science of Ascensions of the Signs of the Zodiac," a commentary on Ptolemy's *Apotelesmatika tetrabiblos*, and a third work on trigonometry, for example, are all lost in their original versions.

Significance

The scholarly career of al-Battani provides an example of the diversity of pre-Islamic sources that contributed to the rise of Islamic science. It also illustrates the importance of such scientists' work in saving traces of pre-Islamic contributions to knowledge during the Dark Ages of European history, when much of the classical heritage of Western civilization was lost. To speak of al-Battani's role as that of an interim transmitter of knowledge, however, would be to miss the essential importance of scientific endeavors in his era. It is clear, for example, that al-Battani was dissatisfied with interpretations offered by his classical and Indian forerunners. By the time his work of reinterpretation was translated for transmission to the European world, it reflected numerous original contributions. Thus, in regard to the reemergence of Western science during the classical revival period of the Renaissance, it can be said that many of the principles on which it was based came from Islamic sources.

The fact that such advances in several fields of "pure" science were actively sponsored by the early Islamic caliphs—themselves assumed to be primary guardians of the religious interests of their realm—is of major significance. In al-Battani's age, knowledge was still recognized as something necessarily derived from syncretic sources. Tolerance for the exploration of different secular scientific traditions did not, however, survive many successive generations. Narrowness of views in the eastern Islamic world a mere century and a half after al-Battani's contributions would make the role of Western translators of Arabic scientific works just as vital to the conservation of cumulative knowledge in world culture as the work of Islamic translators and commentators had been after the end of the classical era. Outstanding figures such as al-Battani, therefore, definitely span world civilizations and reflect values that are universal. These are easily recognized as such beyond the borders of their chronological time or geographic zone.

Byron D. Cannon

Further Reading

Anawati, G. "Science." In *The Cambridge History of Islam*, edited by P. M. Holt, Ann K. S. Lambton, and Bernard Lewis. 1970. Reprint. Cambridge, England: Cambridge University Press, 1977. This chapter places the general field of science within the overall framework of Islamic civilization and explores interrelationships between religious and scientific attitudes toward knowledge and how each of the two domains was affected by developments in the other. The body of the chapter consists of a field-by-field review of the most important Islamic accomplishments, including advances in arithmetic, algebra, geometry, trigonometry, optics, mechanics, astronomy and astrology, and medicine.

Bakar, Osman. *The History and Philosophy of Islamic Science*. Cambridge, England: Islamic Texts Society, 1999. Discusses questions of methodology, doubt, spirituality and scientific knowledge, the philosophy of Islamic medicine, and how Islamic science influenced medieval Christian views of the natural world.

Bell, E. T. *The Development of Mathematics*. New York: McGraw-Hill, 1945. This comprehensive work on mathematics begins with a historical review of the field from the first-known texts through each successive stage of discovery, ending at the mid-point of the twentieth century. Includes the chapter "Detour

Through India, Arabia, and Spain, A.D. 400-1300." Covers the specific contributions of Islamic scholars such as al-Battani and examines topical subsections of the field of mathematics, including geometry, invariance, and others. Bibliographical references.

Cajori, Florian. *A History of Mathematics*. 1931. 5th ed. Providence, R.I.: AMS Chelsea, 2000. This classic work has several important characteristics that still merit mention. It covers standard non-Western mathematical traditions (Hindu and Islamic). The author gives detailed information on individual mathematicians' original findings.

Hogendijk, Jan P., and Abdelhamid I. Sabra, eds. *The Enterprise of Science in Islam: New Perspectives*. Cambridge, Mass.: MIT Press, 2003. A collection surveying the history of Islamic science, including mathematics and astronomy. Illustrations, bibliography, index.

Kennedy, Edward S. *Astronomy and Astrology in the Medieval Islamic World*. Brookfield, Vt.: Ashgate, 1998. Discusses astronomy, astronomers, math, and mathematicians during the Islamic Middle Ages. Includes an index.

Nasr, Seyyed Hossein. *Islamic Science: An Illustrated Study*. London: World of Islam Festival, 1976. Contains an attractive collection of illustrations that bring to life the world of Islamic science. These include intricate miniatures depicting cosmic charts and photographs of astronomical instruments and remains of observatories similar to the ones that would have been used in al-Battani's time. Glossary and bibliography.

_______. *Science and Civilization in Islam*. Cambridge, Mass.: Harvard University Press, 1968. This survey text deals only with scientific endeavors in Islamic civilization, which allows it to examine individual contributions thoroughly. Extensive bibliography.

Sabra, Abdelhamid I. *Optics, Astronomy, and Logic: Studies in Arabic Science and Philosophy*. Brookfield, Vt.: Variorum, 1994. A history of science, specifically optics and astronomy, in the Muslim world of al-Battani's time. Illustrations, index.

Stephenson, F. Richard. *Historical Eclipses and Earth's Rotation*. New York: Cambridge University Press, 1997. Explores "Solar and Lunar Eclipses Recorded in Medieval Arab Chronicles," "Observations of Eclipses by Medieval Arab Astronomers," and "Eclipse Records from Medieval Europe." Illustrations, bibliography, index.

FRIEDRICH WILHELM BESSEL

German astronomer

Bessels contribution to astronomy was to increase greatly the accuracy of measuring the positions of stars by using more advanced instruments and developing methods to account for instrument and observer error. The most famous discovery resulting from these observations was the first accurate determination of the distance to a star.

Born: July 22, 1784; Minden, Brandenburg (now in Germany)
Died: March 17, 1846; Königsberg, Prussia (now Kaliningrad, Russia)

Early Life

Friedrich Wilhelm Bessel (BEHS-ehl) was born to a civil servant and a minister's daughter. One of nine children, he went to the local gymnasium but left after only four years to become a merchant's apprentice. He showed no particular talent at school. At the age of fifteen, he began his unpaid seven-year apprenticeship to a merchant firm in Bremen. He excelled at his accounting job and received a small salary after one year. He spent his spare time teaching himself geography and languages because of his interest in foreign trade. He also learned about ships and practical navigation through self-study. Determining the position of a ship at sea—a long-standing problem in navigation—intrigued him. He therefore began to study astronomy and mathematics to understand the theory behind the existing methods.

Bessel began to make observations of stars on his own, which he was able to compare to observations reported in numerous professional journals of astronomy. One of his earliest tasks was to determine the orbit of Halley's Comet based on several observations of its position. He studied existing methods to determine the easiest way to do this, and he used observations made in 1607 to supplement his own. The precision of his observations and the scrupulous care given to minimizing or correcting for observational errors would characterize his professional work throughout his life. He made the observations, adjusted (or reduced) the 1607 data to make them directly comparable to his own, and submitted the results to Wilhelm Olbers in 1804.

Olbers, a physician and highly esteemed amateur astronomer, was impressed with the agreement between Bessel's observations and Edmond Halley's calculation of the orbit. He urged Bessel to improve it further with more observations. Olbers was impressed enough to recommend Bessel in 1806 for a position as assistant at a private observatory near Bremen. Bessel made further observations of the comet and published the results in 1807 to wide acclaim.

Life's Work

In 1809, Bessel was appointed director of the new observatory at Königsberg, where he remained for the rest of his life. His early fame came from his reduction of the earlier observations of James Bradley. Bradley's measurement of the apparent position of stars had to be corrected for the motion of Earth, the bending of starlight as it passes through the air, and instrument errors. With sufficient care, any observation can be reduced to a universal coordinate system.

While waiting for the construction of the observatory, Bessel worked on reducing Bradley's observations of more than thirty-two hundred stars with the goal of producing a reference system for measuring the positions of other stars. Bessel received the Lalande Prize of the Institute of France for his production of tables of refraction based on these observations. In 1818, he completed thereduction and published the results in his work *Fundamenta astronomiae* (1818; fundamental astronomy). This work provided the most accurate positions of a chosen set of stars. Accurate positions of a few stars are required to form the basis for extremely accurate measurements of positions for all other stars. This work has been said to mark the birth of modern astrometry. Bessel also provided accurate proper motions of many stars. (The so-called fixed stars actually move a very small amount over the centuries. When all perturbing effects are removed, the motion that is left to the star is called its proper motion.)

Bessel's next important contribution was to increase the accuracy of the measurement of stellar positions and motions. In 1820, he determined the position of the vernal equinox with great accuracy. The equinox is employed as the origin of the coordinate system used to record a star's position. He further improved accuracy in his work *Tabulae regiomontanae* (1830; *Refraction Tables*, 1855), in which he published the mean positions for thirty-eight stars for the period 1750-1850. In 1821, he noticed a systematic error in observation that was peculiar to each observer and called it the personal equation. This systematic error was reduced as each observer became more experienced, but it never disappeared. Bessel devised a method to remove the error.

Identifying and measuring the proper motion of starswas crucial in producing an important contribution to astronomy. The slight but periodic variation in proper motion of a few stars was not accountable by considering the motion of Earth or instrumental factors. In making the observations that later appeared in the *Refraction Tables*, Bessel suggested that the variations in proper motion of the stars Sirius and Procyon could be explained by the existence of an as-yet-unseen companion star. More than a century later, the companions to these stars were observed, as well as companions to many others.

Bessel also made important contributions to mathematics. Prior to Bessel, it was common for observers of the heavens to record their data and only later, if ever, reduce that data. Bradley, whose observations Bessel used extensively, carefully noted any possible perturbing effects in his observations. Nevertheless, reduction of the data for the positions of the stars was put aside in favor of recording lunar data. Bessel emphasized the need for the data reduction to be done immediately by the observer. Such reduction required extensive manipulation of complicated equations. In the process of developing ways to remove errors, Bessel noticed that he could use a class of functions that solved problems involving the perturbing influence of one planet on the orbit of another. He systematically investigated and described this class of functions in 1824. These functions, which bear Bessel's name, are not restricted to astronomy: They are used in the solution of a wide variety of problems in physics, mathematics, and engineering.

Another direct benefit of increased accuracy in stellar positions was Bessel's determination of the distance to a star, which is his most important contribution to astronomy. Although many astronomers had earlier claimed to have measured stellar distance using methods based on questionable assumptions, and although two of Bessel's contemporaries also correctly determined such distances, Bessels comprehensive treatment of the data and his high accuracy of observations were convincing to his contemporaries.

As Earth moves in its annual orbit around the sun, the stars appear to move across the dome of the sky. By determining the location of a star at opposite ends of Earth's orbit and using some simple trigonometry, the distance to a star can be measured. However, this so-called parallax (the apparent motion of an object caused by the motion of the observer) was small because the stars are very far away. The parallax had therefore never been measured. Indeed, some opponents of the heliocentric theory, according to which Earth revolves around the sun, used this failure to measure parallax as an argument against the theory. Astronomers had a rough figure for the radius of Earth's orbit and some idea of the extent of the solar system, but there did not seem to be a way to determine stellar distances that did not require the assumption that all stars had the same intrinsic brightness.

Earlier attempts at measuring parallax involved circumventing the problem of the immeasurably small parallax by looking at two stars that appeared to be very near to each other but were of different brightness. It was thought that the dimmer star would be farther away and that observing the relative change of position could lead to a determination of stellar distances. This method did not work, because not all stars are of the same intrinsic brightness. Most stars that appear near to each other are in fact binary stars and really are near each other.

Bessel used a different approach: He assumed that stars with large proper motions are closer than stars with small proper motions. He chose the star known as 61 Cygni, because it had the largest proper motion known. He used a new measuring device called the Fraunhofer heliometer (after Joseph von Fraunhofer, a nineteenth century optician), which was designed to measure the angular diameter of the sun and the planets. Its manner of comparing the images from two objects to determine angular diameter was more accurate than earlier instruments. Using nearby stars for comparison and observing for eighteen months, Bessel was able to measure a parallax of slightly under one-third of a second of arc, which is equal to the width of a dime viewed from twenty miles. From this amount of parallax, Bessel calculated that 61 Cygni was 10.9 light years or seventy trillion miles away. He completed his calculations and published the results in 1838.

The last six years of Bessel's life were marked by deteriorating health, but he managed to complete a number of works before his death from cancer in March, 1846.

Significance

Friedrich Wilhelm Bessel made important contributions to astronomy, mathematics, and geodesy. His work marks the turning point from a concern with planetary, solar, and lunar observations to investigations of the stars. His measurement of the distance to a star is noteworthy because it settled the centuries-old question of whether stars exhibited parallax. The care he took in his observations set much higher standards for the science of astronomy. Bessel's goal was to observe the stars accurately enough to predict their motion and to establish a reference system for their positions. As part of that plan, he developed methods for the careful determination of instrument and observer error, conducted years of observations himself, and developed the mathematical techniques to reduce the data. Bessel's lasting achievement was to raise the science of observing, reducing, and correcting astronomical data to an art.

Roger Sensenbaugh

Further Reading

Clerke, Agnes. *A Popular History of Astronomy During the Nineteenth Century*. New York: Macmillan, 1887. Although written during the period it was supposed to cover, this work has several redeeming qualities. Expressly written for a general audience, the book's language is clear and precise. A valuable record of what near-contemporaries thought of Bessel.

"The Deep Sky." *Astronomy* 31, no. 8 (August, 2003): 65. Describes the special place the star system Cygnus holds in the history of astronomy because of Bessel's measurements of 61 Cygni.

Herrmann, Dieter B. *The History of Astronomy from Herschel to Hertzsprung*. Translated by Kevin Krisciunos. Rev. ed. New York: Cambridge University Press, 1984. Traces the history of astronomy from 1780 to 1930. Written from a Marxist perspective.

Hirshfeld, Alan. *Parallax: The Race to Measure the Cosmos*. New York: W. H. Freeman, 2001. Bessel's measurement of 61 Cygni is recounted in this history of astronomers' attempts to measure the distances to stars and plants by parallax observation.

Hoskin, Michael A. *Stellar Astronomy: Historical Studies*. Chalfont, England: Science History, 1982. A collection of material published in *Journal for the History of Astronomy*, with the addition of some new material. Attempts a synthesis of existing scholarship on the history of stellar (as opposed to planetary) astronomy as of the early 1980s.

Pannekoek, Anton. *A History of Astronomy*. New York: Interscience, 1961. Traces the history of astronomy from antiquity to the present. Part 3, "Astronomy Surveying the Universe," contains information on Bessel and places him in the historical context of nineteenth century astronomy.

Williams, Henry Smith. *The Great Astronomers*. Westport, Conn.: Greenwood Press, 1930. Reprint. New York: Newton, 1932. Book 5 deals with Bessel, among other subjects. Concerns parallax and Bessels contributions to measuring the distance to 61 Cygni. Describes the work of Bessel's contemporaries who measured the parallax of other stars at about the same time.

TYCHO BRAHE

Danish astronomer

Brahe realized early that the existing means for observing and measuring celestial bodies and their motions were inaccurate. His great achievements led to significant improvements in existing instruments, the invention of new instruments, and amazingly accurate observations.

Born: December 14, 1546; Knudstrup Castle, Scandia, Denmark (now in Sweden)
Died: October 24, 1601; Prague, Bohemia (now in the Czech Republic)
Also known as: Tyge Brahe Ottosøn
Areas of achievement: Astronomy, mathematics, science and technology

Early Life

Tycho Brahe (TI-koh BRAH-hee), born Tyge Brahe Ottosøn, was the eldest of ten children born to Otto Brahe. The Brahes were an old and noble family with both Danish and Swedish branches. Tyge's father was privy councillor to the king of Denmark at the time of Tyge's birth and ended as governor of Helsingborg Castle in Scandia—then part of Denmark. Tycho was not reared with his parents and younger siblings. His father's brother Jørgen, who was childless, stole him while he was still a baby. Initial turmoil in the family was stilled when Otto's second son was born. It was therefore in the home of his uncle that Tyge was reared, showing so much early scholarly promise that, in addition to the requisite training for a young nobleman in horseback riding and swordsmanship, he was allowed to learn Latin in the hope that he would become a statesman and counselor to the king.

At thirteen, Tyge entered the University of Copenhagen, head held high above a piped collar and small rapiers by his side. Thirteen was not an unusual age for university entry at the time. He studied, as did most of his colleagues, philosophy, rhetoric, and law. The curriculum was in Latin. His academic career was planned for him: After finishing his studies in Denmark, he would go to one of the more famous German universities and study law, still in preparation for a career in government. The problem was that young Tyge (he took the name Tycho on graduation) was not interested in law. An event that took place on August 21, 1560, when he was nearly fourteen, came to fascinate him so deeply that it, in effect, determined his choice of career. He heard that an eclipse of the sun had been predicted for that day. That the prediction proved to be correct and that the sun was indeed eclipsed seemed to him divine.

It was not considered good form for one of Brahe's social station to become a mere scientist, and his fascination with astronomy was greeted with far less than enthusiasm by his uncle and father. His astronomical studies were performed in secrecy and at night. When he went to Germany, a tutor accompanied him to ensure that he did not stray from his legal studies. The two arrived in Leipzig in 1562, when Brahe was not quite sixteen. Anders Sørensen Vedel, the tutor, kept a rapt eye on his charge, and Brahe studied law by day and reserved his nights for gazing at the stars. He also managed to study mathematics, which would be necessary for him in his astronomical studies, and he met two of the more famous astronomers of his day, who happened to reside in Leipzig—Bartholomaeus Scultetus and Valentin Thau. The instruments he used for his observations were quite crude: a globe, a compass, and a radius.

Life's Work

The tutor eventually became aware of his charge's illicit nightly activities, and the two of them were called home to be under Jørgen Brahe's intense scrutiny. Yet Jørgen died not long afterward, leaving the nineteen-year-old Brahe wealthy and independent. Brahe could embark on his life's work. For a while, out of a sense of duty, he performed his responsibilities as a nobleman and oversaw the Tostrup estate in Scandia that was part of his inheritance. In 1566, however, he decided to make his scientific studies the center of his life and activities.

To the accompaniment of his family's scorn, Brahe moved to Wittenberg and worked with Kaspar Peucer, a then-famous astronomer, until an outbreak of the plague forced his return to Denmark. Later, he went to Augsburg because of its famous instrument makers; he wanted new, more precise, better designed instruments made for his observations. He had a new globe, sextant, and radius made. His fame in the scientific community began to grow.

Brahe concentrated his early studies on the apparent movements of the planets and the fixed stars. His father's death and an appointment as cantor of the Roskilde Cathedral devoured much of his time, but he steadfastly continued his work. His growing fame changed the attitudes of his family and peers toward his work: His uncle Steen Brahe had a lab outfitted for him. His first major breakthrough was the observation of a new star, first seen on November 11, 1572. The star, which he appropriately called Stella Nova in a book entitled *De nova et nullius aevi memoria prius visa stella* (1573; about the new star), appeared in the Cassiopeia constellation. Large and bright, the new star remained visible until 1574. The accuracy of the observations, down to the minute details, caused a sensation in the scientific community, and Brahe was established as a great scholar.

Brahe, at this time a grown man, cut quite a striking figure. Bejeweled and flamboyantly dressed, he was stocky, with reddish-yellow hair combed forward to hide incipient baldness, and he sported a pointed beard and a flowing mustache. When he was young and while in Germany, Brahe had been in a duel, and his opponent sliced off a large piece of his nose, for which Brahe had a substitute made of gold and silver and painted to look natural. He always carried a box with glue and salve.

Brahe's plan was to settle abroad and continue his studies, but the king changed his plans by donating to him the small island of Ven, in the channel between Denmark and Sweden. The position of the island was perfect for his purposes, and Brahe accepted the generous offer. On Ven, he built his famous observatory: the architecturally beautiful Uraniborg (after the muse of astronomy Urania), which contained a chemistry lab, his famous mural quadrant, and observatories in the attic. He also built the smaller, but equally famous, Stellaburg, which, except for a cupola, was built underground. This building contained many observational instruments, including his renowned revolving quadrant. It also had portraits, in the round, of the greats of astronomy: Timocharis, Hipparchus, Ptolemy, Nicolaus Copernicus, and himself.

On the official front, Brahe became royal mathematician and lecturer at the University of Copenhagen. Little is known about his private life. He married a bondwoman, Kirstine, by whom he had eight children, five girls and three boys. Brahe did not reveal this part of his life in his own writings, and his contemporaries restricted themselves to expressing disapproval of this alliance. In his observatories, he continued his work, making surprisingly accurate observations of celestial bodies. He always had a number of students living with him, who helped him in his observations.

By 1582, Brahe had reached a point at which he could propose his own astronomical system. He rejected both the static Ptolemaic system with the sun and planets moving around Earth in individual orbits and the Copernican system, which has Earth and the other planets moving around the sun. Brahe's system is an amalgam. For reasons involving both the laws of physics and the Bible, he could not accept a system that makes Earth simply one of the planets that revolve around the sun. In Brahe's system, Earth is static and the moon and the sun revolve around it, with the other planets revolving around the sun. It remained for Brahe's student Johannes Kepler to reinstate the correct Copernican system reinforced by Brahe's minutely correct measurements and observations.

On Ven, Brahe had his own printing press and published, besides his own works, calendars and horoscopes for the king and other high dignitaries. Like many of his contemporaries, Brahe did not distinguish sharply between astronomy and astrology, but he apparently did not think highly of horoscopes and made them only under duress. Many kings and dignitaries from around Europe visited the island to see the famous observatories.

On the island, Brahe did the bulk of his scientific work. He made accurate observations of the sun, moon, and planets. Many scholars find that his greatest achievement, besides his introduction of the use of transversals on the graduated arcs of astronomical instruments and his improvements of existing instruments (such as the equatorial armillae, which are spheres used to establish differences in longitude and latitude), was his catalog of fixed stars, which stood until such improved instruments as telescopes and clocks of precision came into use.

Unfortunately, Brahe did not adhere to his scientific studies. As he grew older, his idiosyncrasies became more obvious and he became involved in some petty suits that alienated the king, who had been one of his staunchest supporters. Brahe's intransigence finally caused the king to confiscate land that had been bequeathed to him, leaving Brahe without an adequate source of income.

Finally, in July, 1597, Brahe left native shores and moved to Rostock, Germany. He sent a submissive letter to King Christian IV, in which he asked the king to take him back into his good graces. The letter elicited a direct and angry response from the king, who said that until Brahe came to his senses, admitted his faults, and promised to do as he was told, he should not return.

Brahe, determined not to give in, decided to find a new mentor. He approached Emperor Rudolf II in Prague. Rudolf had a reputation as a patron of the sciences and indeed took Brahe and his collaborator Kepler under his wing. The two famous astronomers had, at times, a stormy relationship, and, after several years, Kepler had a nervous breakdown and left Prague. Brahe died on October 24, 1601, in Prague.

Significance

Tycho Brahe was a transitional figure in the history of astronomy. His theoretical work was flawed and actually a step back from the work of Copernicus. His great achievements were in the areas of practical and spherical astronomy. He devised new and more sophisticated instruments for observations and recorded an astounding body of observations that represented a quantum leap forward in knowledge about the movements and relative positions of celestial bodies.

Brahe's observatories represented the state of the art in sixteenth century astronomy. Here he gazed at and recorded the stars, made mathematical computations, and had his most famous instruments built and installed: three equatorial armillae; a mural quadrant, which he used to determine time; and sextants with transversals on the graduated arc and improved sights that allowed for pointing the instrument with great precision to measure distances and angles.

Brahe's legacy, which has made him, in one biographer's somewhat hyperbolic phrase, "a king among astronomers," is his large body of accurate observations and measurements performed by means of instruments and methods devised by him.

Per Schelde

Further Reading

Christianson, John Robert. *On Tycho's Island: Tycho Brahe and His Assistants, 1570-1601*. New York: Cambridge University Press, 2000. Extensive study of Brahe's life and accomplishments in many fields, including philosophy, chemistry, poetry, and reform. The book focuses especially on the vital intellectual community he created, and it provides short biographies of more than twenty members of that community. Includes illustrations, maps, bibliographic references, and index.

_______. *Tycho Brahe: A Picture of Scientific Life and Work in the Sixteenth Century*. Reprint. New York: Dover, 1963. The most detailed work on Brahe in terms of his work and studies. Early in the book, Dreyer sets the general scientific and astronomical stage Brahe was to enter. While somewhat technical, the book gives a thorough and minute description of Brahe's instruments and observations.

Ferguson, Kitty. *Tycho and Kepler: The Unlikely Partnership That Forever Changed Our Understanding of the Heavens*. New York: Walker, 2002. Tells the story of the relationship between Brahe's work and Kepler's, as well as the relationship between the two men. Argues that neither would have achieved fame or lasting effects on science without each other's aid. Includes eight pages of photographic plates, illustrations, maps, bibliographic references, and index.

Gade, John Allyne. *The Life and Times of Tycho Brahe*. Princeton, N.J.: Princeton University Press, 1947. Reprint. New York: Greenwood Press, 1969. Gade gives the social and political backdrop to Brahe's life and work. His emphasis is not so much on technical descriptions as on Brahe the person and the community member. Gade writes amusingly of Brahe's childhood and youth and gives a fairly complex psychological profile of the adult scientist and nobleman. The most personal portrait of Brahe extant.

Gray, R. A. "Life and Work of Tycho Brahe." *Royal Astronomical Society of Canada Journal* 17 (1923). Starts with a careful statement of the Ptolemaic and other theories current before the advent of Brahe. Lists among Brahe's achievements his statement that comets are not, as previously believed, within Earth's atmosphere. Also mentions Brahe's improvements on existing instruments.

Thoren, Victor E., with John R. Christianson. *The Lord of Uraniborg: A Biography of Tycho Brahe*. New York: Cambridge University Press, 1990. Massive and comprehensive biography that attempts to reevaluate and reinterpret nearly every aspect of Brahe's career and contribution to science. Includes illustrations, bibliographic references, index.

BRAHMAGUPTA

Indian mathematician

Brahmagupta wrote Brahmasphuzasiddhānta, a book in verse expounding a complex system of astronomy and containing two chapters on arithmetic, algebra, and geometry. His work on indeterminate equations and the introduction of negative numbers greatly influenced the development of science in both India and Arabia.

Born: c. 598; Bhillamala, Rajputana (now Bhinmal, India)
Died: c. 660; possibly Ujjain, Kingdom of Magadha (now in India)

Early Life

The Hindu astronomer and mathematician Brahmagupta (brah-mah-GEWP-tah) was born to a man named Jishnugupta from the town of Bhillamala. The suffix *-gupta* may indicate that the family belonged to the Vaiśya caste (composed mostly of farmers and merchants).

In contrast to his predecessor, Āryabhata the Elder (c. 476-c. 550), who lived in relative obscurity at Kusumapura (modern-day Patna, Bihar), Brahmagupta had the opportunity to live, study, and teach in Ujjain, a town in the state of Gwalior, Central India. Ujjain was then the center of Hindu mathematics and astronomy and had the best observatory in India. At Ujjain, Brahmagupta also had access to the writings of many great scientists who came before him, including Hero of Alexandria (fl. 62-late first century C.E.), Ptolemy (c. 100-178 C.E.), Diophantus (fl. c. 250 C.E.), and Āryabhata the Elder. Brahmagupta later drew heavily on these sources in his own writings, often correcting their errors. For example, he corrected Āryabhata's mistake regarding the formulas for the surface areas and volumes of the pyramid and cone. Brahmagupta even borrowed mathematical problems, including one calling for the calculation of the position of a break in a bamboo pole. This problem had first appeared in the Chinese text *Jiuzhang shuanshu* (c. 50 B.C.E.-100 C.E.; arithmetic in nine sections), the authorship and date of which are uncertain.

Another influence of Ujjain was on Brahmagupta's style of writing. Like other Hindu scientists, including Āryabhata, he wrote his mathematical texts as poetry. The Indian practice was to clothe all arithmetical problems, especially those in schoolbooks, in poetic garb, fashioning them into puzzles that served as a popular amusement. Brahmagupta wrote that his mathematical problems were undertaken only for pleasure and that a wise man could invent a thousand more or solve those presented by others, thereby eclipsing their brilliance, just as the sun eclipses the other stars in the sky.

At Ujjain, the thirty-year-old Brahmagupta completed his masterwork, the *Brahmasphuzasiddhānta* (c. 628; the improved astronomical system of Brahma). The date of this work has been determined by consulting both commentary from later Hindu scholars and, appropriately, astronomical data.

Life's Work

The first ten chapters of Brahmagupta's *Brahmasphuzasiddhānta* deal with various astronomical issues, including the mean and true longitudes of the planets, diurnal motion, lunar and solar eclipses, heliacal risings and settings, the lunar crescent and "shadow," conjunctions of the planets, and their conjunctions with the stars. The following thirteen chapters take up an examination of previous work on astronomy (including Āryabhata's), additions and problems (and their solutions) supplementing six of the earlier chapters, mathematics, the gnomon, meters, the sphere, instruments, and measurements. The work's twenty-fourth and final chapter summarizes the principles of Brahmagupta's astronomical system in a compendious treatise on astronomical spheres. (Some manuscripts include an additional chapter containing tables.) All but two of the chapters deal with astronomy, but scholars have chosen those two chapters, 12 and 18, which deal with algebra and mathematics, to study most intensely.

Although Brahmagupta studied mathematics only for its applicability to astronomy and considered knowledge of the rules of arithmetic a prerequisite to be a *ganaca* (a student of astronomy), most scholars in the ages since he lived have studied his mathematics more closely than his astronomy. Of particular interest is his work on indeterminate equations, building on the work of both Diophantus and Āryabhata. Brahmagupta's work, along with that of Bhāskara II (1114-c. 1185), solved the so-called Pell equation, $y^2 = ax^2 + 1$, where a is a nonsquare integer. Brahmagupta showed that from one solution where x, y, and xy do not equal zero, a general formula indicating an infinite number of solutions could be derived. Brahmagupta also stated that the equation $y^2 = ax^2 - 1$ could not be solved with integral values of x and y unless a was equal to the sum of the squares of any two integers. Brahmagupta's work on these equations, with additions by Bhāskara, is highly regarded because it was not for several centuries that another mathematician, namely Joseph-Louis Lagrange (1736-1813), could completely work out the Pell equation.

Brahmagupta also studied indeterminate equations of the first order, such as this one: Two ascetics live on top of a hill of h units of height, whose base is mh units away from a nearby town. One ascetic descends the hill and walks directly to the town. The other, being a wizard, flies straight up a certain distance, x, then proceeds in a straight line toward the town. If the distance traveled by each ascetic is the same, and h is 12 and mh is 48, find x. The solution comes from the formula $x = mh/(m + 2)$, or in this case, $x = 8$.

Brahmagupta's work on the geometry of quadrilaterals, which was probably inspired by his studies of

Ptolemy and Hero, is also a landmark in the history of Hindu mathematics. Brahmagupta found the formulas, for the first time, for the diagonals (defined as m and n) of a quadrilateral having sides of length a, b, c, and d and opposite angles of A and B, and C and D. He calculated the diagonals thus:

$$m^2 = (ab + cd)(ac + bd)/(ad + bc) \text{ and}$$
$$n^2 = (ac + bd)(ad + bc)/(ab + cd).$$

These formulas were later studied by Bhāskara, who, failing to understand that they applied only to quadrilaterals inscribed in a circle, incorrectly pronounced them unsound. Brahmagupta also figured that, if a, b, c, A, B, and C are positive integers such that $a^2 + b^2 = c^2$ and $A^2 + B^2 = C^2$, then the cyclic quadrilateral having consecutive sides aC, cB, bC, and cA (which came to be called a Brahmagupta trapezium) has rational area and diagonals, and the diagonals are perpendicular to each other. These formulas are most remarkable; nothing like them had previously appeared in Hindu geometry.

Brahmagupta borrowed from Hero of Alexandria the formula for the triangular area, but he brilliantly extended Hero's formula to work with quadrilaterals that can be inscribed within circles. This idea was later built on by the ninth century Hindu mathematician Mahāvīra and was much admired by later commentators. Brahmagupta's other advances in mathematics included proving the Pythagorean theory of the right triangle, deriving formulas for the areas of a square and a triangle inscribed in a circle, and showing that a rectangle whose sides were the radius and semiperimeter of a circle had the same area as that circle.

Although he is now remembered mostly for his advances in mathematics and his influence on the mathematical work of later Hindus such as Mahāvīra and Bhāskara, Brahmagupta considered himself primarily an astronomer. Almost every Hindu commentator on astronomy discusses his work. Indeed, some of his work in astronomy is quite admirable. He provided fairly accurate figures for the circumference of Earth and the length of the calendar year. Brahmagupta gives a figure different from Āryabhata's for the circumference of Earth: 5,000 *yojanas*. Assuming that Brahmagupta's *yojana* was a short league of about 4.5 miles, that would convert his figure to 22,500 miles, which is not too far off the mark. He also tried to correct Āryabhata's computation for the length of the year, which was 365 days, 15 *ghati*, 31 *pala*, and 15 *vipala*, or 365 days, 6 hours, 12 minutes, and 30 seconds. His own figure was slightly more accurate: 365 days, 15 *ghati*, 30 *pala*, 22 *vipala*, and 30 *pratipala* (365 days, 6 hours, 12 minutes, and 9.0 seconds).

Much of his astronomy, however, is quite erroneous. Like many Hindu scientists of the time, Brahmagupta was vehemently opposed to Āryabhata's ideas that Earth revolved around the Sun and spun on its axis. Why then, Brahmagupta asked, do not the lofty bodies fall down to Earth? He also questioned Āryabhata's theory of an aerial fluid that causes Earth to rotate.

Significance

Although Brahmagupta greatly extended the work of many preceding mathematicians and presented numerous valid theories of his own, it must be acknowledged that he did make some serious scientific errors. In addition to denying Āryabhata's theories of the place of Earth in the solar system, he gave a faulty formula for the area of an equilateral triangle. In his studies on the circle, he alternately used 3 and the square root of 10 as values for π.

Yet Brahmagupta's importance as a scientist must have been recognized during his lifetime, because he was accused of propagating scientific falsehoods to please the priests and the ignorant commonfolk. The priests were particularly opposed to the ideas that Earth was round and that it rotated around the Sun. Perhaps Brahmagupta had lied to avoid the fate of Socrates (c. 470-399 B.C.E.).

Despite these accusations, at least two of Brahmagupta's algebraic formulations, although originally devised for use in astronomy, became widely used by Hindu traders. Of particular practical use was his rule of three, in which the Argument, the Fruit, and the Requisition are the names of the terms. The first and last terms have to be similar. The Requisition multiplied by the Fruit and divided by the Argument yielded the Produce.

Brahmagupta also introduced the use of negative numbers, which he used to unify three of Diophantus's quadratic equations under a general equation. These negative numbers were especially useful to merchants in representing debts, along with positive numbers, which represented assets. Another advance in mathematics that the merchants must have found helpful was Brahmagupta's work on interest rates.

By 700, Hindu merchants had introduced Brahmagupta's mathematics to the Arabs, with whom they carried on a high volume of trade. In 772, a table of sines from Brahmagupta—which, incidentally, was probably based on work by Āryabhata—reached the Abbāsid caliph al-Mansūr, and it was ordered to be translated into Arabic. The entirety of the *Brahmasphuzasiddhānta* was translated into Arabic by 775, around the time works by other Greek and Hindu mathematicians were being translated by Arab scholars. Together, these works would greatly influence the nascent Arabic mathematics, with Brahmagupta's greatest contributions coming in the study of negative numbers and indeterminate equations.

Frank Wu

Further Reading

Ball, W. W. Rouse. *A Short Account of the History of Mathematics*. 4th ed. London: Macmillan, 1908. A thorough overview of the history of mathematics, with a section on Brahmagupta and his work on quadratic equations, right triangles, and algebra, plus scattered information on his later influence on Hindu and Arab mathematicians.

Cajori, Florian. *A History of Mathematics*. 5th ed. Providence, R.I.: AMS Chelsea, 2000. Gives the solution to Brahmagupta's broken bamboo problem, plus formulas for Brahmagupta's work on triangles and quadrilaterals.

Eves, Howard. *An Introduction to the History of Mathematics*. 6th ed. Philadelphia: Saunders College, 1990. Includes information on Brahmagupta's studies on indeterminate equations, the Pell equation, cyclic quadrilaterals, and the rule of three, along with a discussion of his place in the history of mathematics. Some problems (with solutions) based on his formula for the cyclic quadrilateral are included.

Joseph, George Gheverghese. *The Crest of the Peacock: The Non-European Roots of Mathematics*. Rev. ed. Princeton, N.J.: Princeton Unversity Press, 2000. Joseph examines the history of mathematics in cultures throughout the world, including India. Its wide coverage places India within the greater scope of mathematical development. Bibliography and indexes.

Lakshmikantham, V., and S. Leela. *The Origin of Mathematics*. Lanham, Md.: University Press of America, 2000. Lakshmikantham argues that the importance of the early Indian mathematicians has been underestimated. Bibliography and index.

Prakash, Satya. *A Critical Study of Brahmagupta and His Works*. New Delhi, India: Indian Institute of Astronomical and Sanskrit Research, 1968. A comprehensive study of Brahmagupta, his works, his sources, and the influence of his work on later writers. Contains an extensive bibliography.

Puttaswamy, T. K. "The Mathematical Accomplishment of Ancient Indian Mathematicians." In *Mathematics Across Cultures: The History of Non-Western Mathematics*, edited by Helaine Selin. Boston: Kluwer Academic, 2000. Examines the early Indian mathematicians and their importance.

GIORDANO BRUNO

Italian philosopher and cosmologist

With his daring and speculative theories in cosmology and philosophy, Bruno anticipated many of the achievements of modern science, but his stubborn personality and arcane interests brought him into inevitable conflict with the authorities of his time.

Born: 1548; Nola, near Naples (now in Italy)
Died: February 17, 1600; Rome, Papal States (now in Italy)
Also known as: Filippo Bruno (given name); Il Nolano
Areas of achievement: Philosophy, astronomy

Early Life

Giordano Bruno (johr-dah-noh BREW-noh) was the son of Juano Bruno, a professional soldier, and his wife, Fraulissa Savolino. As a child, Bruno was named Filippo; he took the name Giordano when he entered the Dominican Order. He was sometimes known as "the Nolan," after the town of his birth, and he often referred to himself in this fashion in his works.

From contemporary records and his own writings, Bruno seems to have been a particularly intelligent and impressionable child. He left several accounts of odd, almost visionary experiences in his youth, including an extended, quasi-mystical dialogue with the mountain Vesuvius that first revealed to him the deceptiveness of appearances and the relativity of all material things. These were to become two dominant themes in his philosophy.

As a youth, Bruno was sent to Naples, where he attended the Studium Generale, concentrating in the humanities, logic, and dialectic. It is clear that Bruno had a thorough grounding in Aristotle and his philosophy and also was well acquainted with the works of Plato and the writings of the Neoplatonists, who were then creating considerable intellectual activity and controversy, especially in Italy.

In 1565, when Bruno was seventeen, he entered the Dominican Order, moving within the walls of the monastery of San Domenico in Naples. There he took the name Giordano. Bruno's decision to enter the Dominican Order is puzzling, for in retrospect it clearly stands as the major mistake in his often-turbulent life. Although he was well suited for the intellectual studies of the Dominicans, he was quite unfit for the accompanying intellectual discipline and submission required for the monastic and clerical life. His thoughts were too wide-ranging and innovative to be restrained within traditional confines, a situation that eventually placed him in mortal conflict with the Church.

Bruno spent eleven years in the monastery of San Domenico. He studied Saint Thomas Aquinas, Aristotle, and other traditional figures, but at the same time was reading in the mystical doctrines of the Neoplatonists, the new works of Desiderius Erasmus, and other reformers and seems to have become suspiciously well acquainted with the works of heretics such as Arius. These unorthodox diversions brought him into conflict with the Dominican authorities, and reports were made that Bruno was defending the Arian heresy. Arius had taught that God the Father and God the Son were not the same in essence. When the Dominicans learned that Bruno was suspected of defending Arianism, charges were prepared against him. Learning of this, he fled the monastery in 1576. He was twenty-eight, and he would spend the rest of his life in exile or in prison.

Giordano Bruno. (Library of Congress)

Life's Work

When Bruno fled the monastery, he embarked on twenty-one years of wandering throughout Europe. Many of his stops lasted merely a matter of months, and the most productive, for only three years. Controversy and conflict dogged him on his travels—much of it a result of not only his daring and speculative thought but also his unrestrained attacks on those who opposed him in any degree and his innate lack of common sense or practical judgment. Employment was difficult, and income was insufficient and insecure. Yet, during this period, Bruno wrote and published an enormous body of work whose content far outpaced even the most advanced thinkers of his time.

Bruno's first extended sojourn was in Geneva. There, safe from the power of the Church, he soon plunged into local intellectual conflicts. In 1579, he published a scathing attack on Antoine de la Faye, a noted professor of philosophy at the University of Geneva. Bruno's assault was more than an academic exercise, for he seemed to undermine de la Faye's theories, which were the basis for the quasi-theological government of Geneva. Bruno, the renegade Dominican on the run, had put himself in disfavor with the Calvinists of Switzerland. He was arrested, then released; he soon left Geneva, moving first to Lyons, then to Toulouse, France. In 1581, Bruno went to Paris, where he found his first real success. He lectured on his own techniques of memory, and the results were so impressive that King Henry III summoned Bruno to court to explain his methods. As a result, the king appointed Bruno to the College de France. Bruno held the post for two years, lecturing on philosophy and natural science and publishing a number of books, many of them on his art of memory.

Still, he managed to alienate many fellow professors and intellectuals in Paris. Some were outraged by his arrogant and self-proclaimed superiority, while the more conventional were troubled by his unorthodox views and desertion of his monastic vows. In 1583, Bruno left for London, with a letter of recommendation to the French ambassador Michel de Castelnau.

The London period, from 1583 through 1585, was the most productive of Bruno's career. Perhaps he was stimulated by the intellectual climate of England, for not only did he deliver a series of lectures at Oxford, explaining the Copernican theory, but also he had among his acquaintances men such as Sir Philip Sidney, Sir Walter Ralegh, and Sir Fulke Greville, noted figures of the English Renaissance. In 1584, Bruno produced a series of six dialogues expounding his philosophy; three of these dealt with cosmological issues and three with moral topics.

In *La Cena de le Ceneri* (1584; *The Ash Wednesday Supper*, 1975), Bruno laid the foundation for his scientific theories. He began with the view of Nicolaus Copernicus that the Sun, rather than the Earth, was the center of the solar system. Bruno recognized that the Sun was itself a star, and he concluded that other stars must have their attendant planets circling them. He came to the conclusion that the universe was infinite, and that it therefore contained an infinite number of worlds, each world capable of having intelligent life on it. Such a theory ran counter to the traditions of both the Catholic Church and the newer Protestant faiths.

Bruno continued the development of his theories in *De l'infinito universo e mondi* (1584; *On the Infinite Universe and Worlds*, 1950). He systematically criticized the prevailing Aristotelian cosmology, and in its place put forth a precursor of the modern theory of relativity later developed by Albert Einstein. Bruno maintained that sensory knowledge could never be absolute, but only relative, and it is this relativity that misleads humans in their attempts to understand the universe. Human perceptions are incapable of truly and completely comprehending the universe, and that universe itself can be accurately comprehended only as a total unity, rather than in isolated parts. Therefore, neither senses nor imagination can be fully trusted, but only reason, which allows humans to penetrate to the divine essence of creation.

Bruno also developed a theory that the universe was composed of "minima," extremely small particles much like the atoms proposed by the ancient Roman philosopher Lucretius. Like Lucretius, Bruno thought that certain motions and events were inevitable and that the universe develops inexorably out of inherent necessity. In order to resolve the conflict between this deterministic view and free will, Bruno postulated that the universe itself was divine; he projected a universal pantheism in which the Creator manifests himself through and within creation.

Finally, Bruno resolved the difficulty of the relationship of human beings to God, of the finite to the infinite, of ignorance to knowledge. These were long-standing puzzles to theologians and philosophers, for it seemed impossible that the limited human mind could comprehend or understand the perfect and infinite attributes of divinity. Bruno believed that there was an identity of opposites at work in which the essential elements of creation and divinity are found in all parts of the universe. Opposition is only relative and illusory; on the most fundamental level, everything is the same, and everything is therefore divine.

In 1585, Castelnau was recalled to Paris, and Bruno, left without a patron, was forced to leave England. For the next six years, he wandered through Europe, accepting and losing posts at a number of universities in Germany and the Holy Roman Empire. He continued to write and publish prolifically, including his special area of memory, and, in the fall of 1591, he received an invitation from a

Bruno on Earth's Recurrent Cycle of Division, Expansion, and Re-Formation

Giordano Bruno advanced the heliocentric theory of Nicolaus Copernicus, believing in an infinite universe with an infinite number of worlds, and thus theorized the expansive and everlasting nature of everything that exists. To expand and re-form, all things, including Earth, "move" so that they may renew themselves.

It (the earth) moves that it may renew itself and be born again, for it cannot endure for ever in the same form. For those things which cannot be eternal as individuals . . . are eternal as a species; and substances which cannot be everlasting under the same aspect, change themselves into other appearances. For the material and substance of things is incorruptible and must in all its parts pass through all forms. . . . Therefore, since death and dissolution are unfitted to the whole mass of which this globe, this star, consists, and complete annihilation is impossible to all nature, the earth changes all its parts from time to time and in a certain order and so renews itself. . . . And we ourselves and the things pertaining to us come and go, pass and repass. . . . And nothing is of itself eternal, save the substance and material of which it is made, and this is in constant mutation.

Source: From the 1584 work *La Cena de le Ceneri*, translated as *The Ash Wednesday Supper* (1975), Giordano Bruno, quoted in *Giordano Bruno and the Hermetic Tradition*, by Frances A. Yates (Chicago: University of Chicago Press, 1979), p. 242.

Venetian nobleman, Zuane Mocenigo, to come to Venice and teach him the art of memory.

Bruno accepted, believing that he would be safe in Venice, which was at that time a fairly liberal and independent state that carefully guarded its freedom from the Papacy. There was a dispute between Bruno and his patron, however—apparently the nobleman believed that he was being cheated and that Bruno planned to flee to Germany—and on May 23, 1592, Bruno was arrested by the Venetian Inquisition. He was questioned through September, but no decision was made.

On February 27, 1593, Bruno was delivered into the hands of the Roman Inquisition, and for the next seven years he was held in prison, repeatedly questioned and examined, and urged to recant his heresies and confess his sins. Bruno tried to play a crafty game, willing to admit minor infractions but pretending not to comprehend how his cosmological and philosophical writings could run counter to the teachings of the Church. Finally, in February, 1600, the Inquisition found him guilty and delivered him to the secular authorities for punishment. When Bruno heard the decision, he replied, "Perhaps you who pronounce my sentence are in greater fear than I who receive it." On Saturday, February 17, 1600, Bruno was burned in the Square of Flowers in Rome.

Significance

Giordano Bruno was a philosopher of great insight and imagination, yet a thinker who could link science to magic and yoke philosophical understanding to mnemonic tricks. He was poised amid the thought and traditions of the Church, the mystical teachings of the Neoplatonists, and the rapid advances of the sciences, especially astronomy. From the combination of these three traditions, he forged a new and highly individual vision of the cosmos and humankind's place in it.

Bruno's influence was recognized by scientists and Humanists in the years following his death. Scientists, even to modern times, admire the startling insights that he drew concerning the infinite number of worlds in an infinite universe. Bruno's early recognition of the concept of relativity and the place it must play in humanity's conception and understanding of the universe is also a prime legacy that Bruno left to science. Humanists of the period were profoundly influenced by his insistence on the need for tolerance in matters of religion and belief. Perhaps because Bruno himself was so often a victim of the intolerance of the age, he was especially eloquent in his plea for patience and understanding.

Finally, Bruno combined the sense of infinite expansion and relativity of all things with a new approach to human knowledge and culture. He refused to divide the world into the sacred and the profane, the Christian and the heathen, the orthodox and the heretic. Instead, he saw human life and culture as a single strand and the universe as a divine manifestation that carried with it all knowledge and truth. To Bruno, the cosmos was God's creation and therefore all good, and humanity's role was not to judge but to understand.

Michael Witkoski

Further Reading

Boulting, William. *Giordano Bruno, His Life, Thought, and Martyrdom*. New York: E. P. Dutton, 1916. Reprint. Freeport, N.Y.: Books for Libraries Press, 1972. As indicated by its title, this biography is a highly favorable account of Bruno's life and thought. On the whole, it presents his philosophical and scientific views in a fair and unbiased light.

de Santillana, Giorgia. *The Age of Adventure: The Renaissance Philosophers*. Boston: Houghton Mifflin, 1957. An introductory survey of Bruno and his work, with particular attention paid to his influence on later scientists and writers.

Feingold, Mordechai. "The Occult Tradition in the English Universities of the Renaissance: A Reassessment." In *Occult and Scientific Mentalities in the Renaissance*, edited by Brian Vickers. Cambridge, England: Cambridge University Press, 1984. An enlightening study of Bruno's 1584 visit to Oxford and the state of learning at that time, with special emphasis on which areas of knowledge were believed to be beyond the boundary of conventions.

Gatti, Hilary. *Giordano Bruno and Renaissance Science*. Ithaca, N.Y.: Cornell University Press, 1999. An explicit attempt to rescue Bruno's reputation as an important contributor to the development of modern science by arguing against seeing him solely as a hermetic philosopher. Compares Bruno's cosmology to those of Copernicus and Kepler; analyzes the scientific import of his overall methodology and worldview. Includes illustrations, bibliographic references, index.

_______, ed. *Giordano Bruno: Philosopher of the Renaissance*. Aldershot, Hampshire, England: Ashgate, 2002. Anthology of eighteen essays about Bruno's life, philosophy, and legacy, organized into sections on his relationship to Italian culture, his experiences and work in England, major philosophical themes in his work, and his historical function as both a receiver of earlier traditions and a source of influence on later thinkers. Includes illustrations and index.

Kristeller, Paul Oskar. *Renaissance Thought and Its Sources*. New York: Columbia University Press, 1979. Does not provide an extended treatment of Bruno as an individual but is an excellent source for understanding the intellectual climate of his time and how it developed.

Singer, Dorothea. *Giordano Bruno: His Life and Thought*. New York: Henry Schuman, 1950. A sympathetic but generally unbiased biography of Bruno, with emphasis on his thought and theory. The volume contains several helpful appendices and an excellent annotated translation of the dialogue *On the Infinite Universe and Worlds*.

White, Michael. *The Pope and the Heretic: The True Story of Giordano Bruno, the Man Who Dared to Defy the Roman Inquisition*. New York: William Morrow, 2002. One of the few Bruno biographies targeted at general readers. Details the social setting in which Bruno lived and the political and religious structures that condemned him to death. Provides a useful account of his influence on later thinkers. Includes bibliographic references and index.

Yates, Frances. *Giordano Bruno and the Hermetic Tradition*. Chicago: University of Chicago Press, 1964. By one of the most distinguished scholars of Renaissance thought. Yates's account of Bruno's place is invaluable. Particularly good in situating him within the confines of a broad philosophical stream.

_______. *Lull and Bruno*. Boston: Routledge & Kegan Paul, 1982. The section "Essays on Giordano Bruno in England" is particularly valuable for its studies of Bruno's lectures on Copernicus at Oxford and his views of religion and the established church.

ANNIE JUMP CANNON

American astronomer

In her work at the Harvard College Observatory, Cannon cataloged stars according to their spectral class into the Draper Catalogue and discovered several new variable stars and novas.

Born: December 11, 1863; Dover, Delaware
Died: April 13, 1941; Cambridge, Massachusetts
Area of achievement: Astronomy

Early Life

Annie Jump Cannon was the only daughter and eldest of three children born to Wilson Lee Cannon and Mary Elizabeth Jump Cannon. The Cannon family also included a half brother and three half sisters, children of her father's first marriage. He was a wealthy merchant and shipbuilder of Scottish descent and was a state senator. Annie's mother was descended from seventeenth century Maryland settlers.

Cannon's family home was prosperous and cultured. She was educated in the Dover public school and in a private academy. Her mother had become interested in stars as a schoolgirl herself and shared her knowledge and interest with Cannon, arranging a makeshift observatory in the attic of the family home. During those years, Cannon learned to recognize constellations and kept a record of her observations. She also was fond of observing the colored spectra produced by prisms on an ornate chandelier in the family home.

After Cannon's graduation from the Wilmington Conference Academy at the age of sixteen, her father sent her to Wellesley College in Massachusetts. One of a group of colleges exclusively for women, Wellesley had opened just five years earlier in 1874. There, under the tutelage of Sarah Frances Whiting, professor of physics and astronomy, Miss Annie (as Wellesley students were thus addressed at that time and as she was known throughout her life) furthered her studies in astronomy and spectroscopy, the study of the spectrum of colors produced when white light passes through a prism.

After her graduation from Wellesley College in 1884, Cannon returned home to Dover and was a popular participant in the social life of that city. She traveled to Spain and made fine photographs on her journey, at a time when photography was not at all a common activity of tourists. She was a gifted pianist, but had early shown signs of hearing impairment that increased as she aged.

In 1893, Cannon's mother died. They had always been very close, and Cannon decided to seek a change in her life. She returned to Wellesley College for a year of further study and to serve as an assistant to Professor Whiting. The following year she enrolled as a special student of astronomy at Radcliffe College, the women's college in Cambridge, Massachusetts. Instruction at Radcliffe paralleled that offered at Harvard College, then an exclusively male institution. As a Radcliffe student, Cannon was allowed access to the Harvard College Observatory, where she began daytime study of photographic plates and evening observation of variable stars.

Annie Jump Cannon. (Library of Congress)

Life's Work

In 1896, Cannon became an assistant at the Harvard College Observatory. The observatory's director, Edward C. Pickering, was well acquainted with Professor Whiting at Wellesley College. Whiting had studied physics with Pickering at the Massachusetts Institute of Technology, where he had served as professor of physics before assuming the directorship of the Harvard College Observatory in 1877. Pickering had advised and assisted Whiting in setting up the Wellesley physics program and made special efforts to encourage the work of Whiting's students.

At the time Cannon joined the observatory staff, the principal work in progress was the cataloging of stars according to their spectral classification through funding from an endowment established in honor of Henry Draper. Draper, a New York physician, was an amateur astronomer particularly interested in stellar spectroscopy. He made the first photograph of the spectrum of a star in 1872, and was assisted in his work by his wife, Mary Anna Draper. After his death in 1882, his widow gave a large sum of money to Harvard to ensure the continuation of this activity.

The introduction of photography and spectroscopy into astronomy in the late nineteenth century made possible a whole new kind of investigation, one that eventually led to the development of astrophysics, the study of the physical conditions and processes prevailing among the stars. By inserting a prism into the light path and replacing the human eye behind the telescope with a camera film, the light from individual stars could be recorded and their spectra analyzed. From photographic observations, astronomers were able to obtain information about stellar temperatures and chemical composition, which varied markedly among the stars.

Pickering was eager to carry on this kind of work. Harvard had the required facilities to do so, not only in Cambridge but also in Peru, where the stars of the southern hemisphere could be photographed, analyzed, and cataloged.

This was the kind of work that Cannon was hired to do and at which she excelled. She joined a group of women working under the immediate supervision of Williamina Fleming. These women were called "computers," since some computation was involved in the analysis of the photographic plates. They worked seven hours a day, six days a week for an average wage of twenty-five cents per hour and were given a month of paid vacation each year. An income of more than $500 per year, while not generous, was sufficient to provide a modest livelihood for a single woman in those days. Cannon also had family money available. During her years at the observatory, her salary increased to $1,200 per year.

Cannon, a keen and careful analyzer, soon realized that the spectra provided new information about variable stars—stars whose brightness varied with time. In 1900, she established a card catalog of all the literature pertaining to variable stars. She also improved on the spectral classification scheme that had been developed by Pickering and Fleming before Cannon joined the staff. In 1911, following Fleming's death, Cannon was named curator of astronomical photographs at the Harvard College Observatory, a post she held until her own death in 1941. Under her care, the photographic collection grew from 200,000 to 500,000 plates.

By 1918, Cannon began to supervise the publication of the Henry Draper Catalogue, which continued until 1924. Spectral classes for some 350,000 stars were included. Still more were added in the Henry Draper Extension, begun in 1925 and continued by others after Cannon's death until 1949. These publications remain the definitive compilation of information on the spectral classes of stars throughout the entire sky.

The wealth of information contained in the Draper Catalogue provided astrophysicists with what they needed to reach an understanding of the evolution of stars and the history of the universe. Because all the information was gathered at one site, the Harvard College Observatory, by one group of analyzers, under the direct supervision of Cannon, the catalog was especially valuable and reliable. The Henry Draper Catalogue became a bible for astrophysicists.

Cannon's work received worldwide recognition and appreciation during her lifetime. Among the honors she received were honorary doctoral degrees from the University of Groningen in 1921 and the University of Oxford in 1925. She was elected to membership in the Royal Astronomical Society of Great Britain, the American Philosophical Society, and the American Academy of Arts and Sciences. In 1938, Harvard University gave her a formal appointment as the William Cranch Bond Astronomer (named in honor of the first director of the Harvard College Observatory, 1839-1859), a singular distinction for a woman at that time.

The male-dominated National Academy of Sciences in Washington, D.C., never elected Cannon to membership, but it did award her its Draper Medal in 1931. She had been given the Nova Medal by the American Association of Variable Star Observers in 1922 and also received the Ellen Richards Prize by the Society to Aid Scientific Research by Women. The latter was a monetary award that Cannon donated to the American Astronomical Society to establish the Annie Jump Cannon Prize. Her namesake prize was to be awarded every three years to a woman astronomer of any nationality who had made a distinguished contribution to astronomy. She stipulated that the award should be a brooch, or pin, that the recipient could wear, rather than a ceremonial medal to be stored in a box.

Cannon retired from the Harvard College Observatory in 1940 and died the following year of heart failure and arteriosclerosis. She was buried in Dover, Delaware.

Significance

Cannon was a product of her particular social class and period in American history. The work she did was made possible by technical developments that had changed the work of astronomers and opened the door to the development of astrophysics. The fact that she was able to accomplish so much for the world of astronomy was due to her own intelligence, dedication and, she said, "patience."

Born into a privileged family, she benefited from American educational opportunities unavailable to previous generations of women anywhere. She was well aware that women, in general, found limited opportunities for participation in society. A believer in the woman suffrage movement, she joined the National Women's Party and was disappointed by the lack of enthusiasm for voting among the general population of women in the United States and elsewhere.

Her lifelong activities centered in Cambridge and in the world of astronomy. She traveled extensively, attending meetings of the International Astronomical Union and working at the Harvard College Observatory site in Peru from where the southern stars could be photographed.

Her visibility in the world of astronomy was an inspiration to other women and led to her being designated, in 1922, America's leading woman scientist by the League of Women Voters. Her establishment of the Annie Jump Cannon Prize provided encouragement to future generations of women astronomers.

There can be little doubt that Cannon's stellar classification scheme, incorporated into the Draper Catalogue, was her principal contribution to world knowledge. With only slight modification it has served several successive generations of astrophysicists. In particular, it was of great value to Henry Norris Russell and Ejnar Hertzsprung in devising their famous Hertzsprung-Russell diagram, which is considered one of the most important astronomical discoveries of the first half of the twentieth century. Cannon's scientific legacy remains a permanent memorial to her dedication and industry.

Katherine R. Sopka

Further Reading

Greenstein, George. *Portraits of Discovery: Profiles in Scientific Genius*. New York: John Wiley, 1998. This book about the sources of scientific genius includes a chapter about Cannon and another female astronomer, Cecila Payne-Gaposchkin.

Jones, Bessie Z., and Lyle Gifford Boyd. *The Harvard College Observatory: The First Four Directorships, 1839-1919*. Cambridge, Mass.: Belknap Press, 1971. Chapter 11, "A Field for Women," contains a biographical sketch of Cannon and a photograph of her in her Oxford University robes. Other chapters provide extensive information on Edward C. Pickering, the Henry Draper Memorial, and on the work of women astronomers at the Harvard College Observatory in considerable technical detail.

Mack, Pamela E. "Straying from Their Orbits: Women in Astronomy in America." In *Women of Science:*

Righting the Record, edited by G. Kass-Simon and Patricia Farnes. Bloomington: Indiana University Press, 1990. This work has particular emphasis on the years up to 1920 and contains a section on Cannon's work at the Harvard College Observatory.

Mozans, H. J. *Woman in Science: With an Introductory Chapter on Woman's Struggle for Things of the Mind.* New York: D. Appleton, 1913. Reprint. Notre Dame, Ind.: University of Notre Dame Press, 1991. A historical source book for women's contributions to science dating to ancient Greece. Chapter 4, "Women in Astronomy," details the work of talented women astronomers, up to and including Cannon. Includes a bibliography of sources. H. J. Mozans is the pseudonym for John A. Zahm.

Ogilvie, Marilyn Bailey. *Women in Science: Antiquity Through the Nineteenth Century: A Biographical Dictionary with Annotated Bibliography.* Cambridge, Mass.: MIT Press, 1986. Contains an essay summarizing Cannon's career and achievements, followed by a list of sources, including obituary notices.

Rossiter, Margaret W. *Women Scientists in America: Struggles and Strategies to 1940.* Baltimore: Johns Hopkins University Press, 1982. A scholarly survey of the opportunities and difficulties encountered by American women scientists and would-be scientists of the nineteenth and early twentieth centuries. Includes specific, referenced examples pertaining to Cannon and her work.

Yost, Edna. *American Women of Science.* Rev. ed. Philadelphia: Lippincott, 1955. An informal biographical account of Cannon and her work, written as part of a series of sketches for high school girls seeking female role models in science.

GIAN DOMENICO CASSINI

Italian astronomer

Cassini accurately determined the rotation periods of Jupiter and Mars, observed a gap in the rings of Saturn, and observed the phases of Venus. He also discovered that Jupiter and Saturn were not spherical but were slightly flattened at their poles. His astronomical measurements, dismissed as inaccurate by Cassini himself, were nevertheless used by Danish astronomer Olaus Rømer in 1676 to calculate the speed of light.

Born: June 8, 1625; Perinaldo, Imperia, Republic of Genoa (now in Italy)
Died: September 14, 1712; Paris, France
Also known as: Jean Dominique Cassini
Areas of achievement: Astronomy, mathematics

Early Life

Gian Domenico Cassini (zhyahn doh-MAY-nee-koh kah-SEE-nee) was born in Perinaldo on the Mediterranean coast, near Nice, France. His father, Jacopo Cassini, was from Tuscany, and his mother was Julia Crovesi. Cassini was raised by an uncle, his mother's brother.

Cassini studied poetry, mathematics, and astronomy at the Jesuit College in Genoa. Initially, Cassini was interested in astrology, but he soon became convinced that astrological predictions were not accurate. Nonetheless, Cassini's knowledge of astrology led to his first professional employment. In 1644, a senator from Bologna, the marquis Cornelio Malvasia, who had a deep interest in astrology and was familiar with Cassini, offered him a position at the new Panzano Observatory in Bologna. Beginning in 1648, Cassini would observe there, using instruments purchased with money obtained by Malvasia. Moving to Bologna was important for Cassini, since he had the opportunity to learn from two outstanding Jesuit scientists: Giovanni Battista Riccioli, an astronomer who made highly detailed telescopic observations of the Moon, and Francesco Maria Grimaldi, a physicist who discovered the diffraction of light.

Life's Work

In 1650, at the age of twenty-five, Cassini became a professor of mathematics and astronomy at the university in Bologna, Italy. While at Bologna, from 1650 to 1671, Cassini made many planetary observations using telescopes built by two contemporary Italians, Eustachio Divini and Giuseppe Campini.

His first serious astronomical observations were of the Sun, and he measured its position with an instrument called a gnomen, a device that used a small hole to allow the Sun's rays to form an image on the floor and, thus, allow Cassini to chart its position accurately. He had this gnomen built to replace an older device that was in the church of San Petronio in Bologna. Cassini was able to determine the precise times of the solstices, that is, the time the Sun reaches its farthest north and farthest south positions in the sky, marking the beginning of winter and summer respectively. In 1652-1653, Cassini observed a comet.

Cassini had initially believed in an Earth-centered solar system. In 1659, however, he presented his own version of an Earth-centered system, in which the Moon and the Sun orbited Earth while the other planets orbited the Sun. He would later come to accept the Copernican model, in which all the planets, including Earth, orbit the Sun.

In July of 1664, using a telescope built by Campini, Cassini observed that Jupiter was not a perfect sphere, as first believed, but instead was flattened at its poles; he also saw that there were bands and spots on the planet. He measured Jupiter's period of rotation by determining how long it took for the large red spot in Jupiter's atmosphere to circle the planet. His value for the rotation period of 9 hours and 56 minutes, published in 1665, is within a few minutes of the best value obtainable with modern instruments.

Cassini observed the moons of Jupiter between 1666 and 1668, and discovered discrepancies in his own measurements that, at first, he attributed to light having a finite speed. However, he appears to have rejected his own idea. In 1676, Danish astronomer Olaus Rømer used Cassini's measurements to calculate the speed of light.

Gian Domenico Cassini. (Library of Congress)

In 1666, he observed surface features on Mars, including Syrtis Major, a feature discovered by Christiaan Huygens in 1659. As with Jupiter, Cassini observed these features as they rotated around the planet and measured the red planet's rotation period. His value of 24 hours and 40 minutes is within 3 minutes of the number accepted in the early twenty-first century.

Cassini also attempted to determine the rotation period of Venus: 23 hours, 20 minutes. However, it is not clear what Cassini observed or what he interpreted as surface features. Modern telescopes show no features on Venus, which is entirely covered by bright clouds, and its rotation period, which can be determined only by mapping surface features with radar that penetrates the clouds, is about 243 days.

Cassini's measurements made him famous in scientific circles throughout Europe, and his work came to the attention of Jean-Baptiste Colbert, the French minister of finance during the reign of King Louis XIV. Colbert had been attempting to strengthen science in France, so, at Colbert's suggestion, the king invited Cassini to Paris to head the new Paris observatory. Cassini arrived in Paris on April 4, 1669, and immediately joined the new Académie Royale des Sciences (Royal Academy of Sciences), which Colbert had founded. When Cassini moved to France, he adopted the French version of his name, Jean Dominique Cassini.

Saturn was a major focus of Cassini's efforts at the Paris observatory. Huygens had discovered Titan, the largest moon of Saturn in 1655. Cassini discovered two other moons, Iapetus in 1671 and Rhea in 1672. In 1675, he recognized that Saturn's ring, discovered by Huygens in 1656, was divided into two parts, separated by a dark gap that is now known as the Cassini Division. In 1677, Cassini demonstrated that Saturn was not a perfect sphere but was flattened at its poles, and he discovered two more moons, Dione and Thetys, in 1684. In 1705, he suggested that Saturn's ring might not be a solid disk, but rather a swarm of objects moving around the planet, objects so small that they could not be seen individually.

Cassini also observed several comets between 1672 and 1707. From December 5 to December 23, 1690, Cassini observed a feature that appeared in the atmosphere of Jupiter. That feature is similar to the features observed in 1994, when more than twenty observable fragments of Comet Shoemaker-Levy 9 impacted Jupiter. Amateur astronomer Isshi Tabe and professional astronomer Junichi Watanabe, both Japanese, believe that Cassini's drawings show that he observed the effects of a similar comet impact in 1690. Watanabe and colleagues published the findings in a 1997 journal article.

During his years at the Paris observatory, Cassini served as the organizer of a renowned group of astronomers called the Paris School, and it included Jean Picard, Huygens, Rømer, Giacomo Felippo Maraldi, and Philippe de La Hire. Cassini was also trained in engineering and published several manuscripts on flood control. He served as inspector of water and waterways and as superintendent of the fortifications of Fort Urban.

In 1673, Cassini became a naturalized French citizen and married Geneviève Delaître, the daughter of an adviser to the king. They had two sons. The younger son, Jacques Cassini (1677-1756), followed in his father's footsteps, taking over his duties as head of the Paris observatory in 1712. Cassini's grandson César-François Cassini de Thury (1714-1784) and his great-grandson, Jacques-Dominique de Cassini (1748-1845) also became noted astronomers. By 1711, Cassini was blind, and he died in Paris on September 14, 1712.

Significance

Cassini's improvement of the gnomen in the church of St. Petronius allowed him to determine precisely the times of the solstices. His measurements were influential in the adoption of the Gregorian calendar in the Protestant countries in 1700, by which time the Julian calendar was eleven days behind the actual beginning of each season. The Gregorian calendar, which corrected errors in the earlier Julian calendar, had been introduced in the Catholic countries of Europe in 1582.

Civil and Jesuit missionaries came to the Paris observatory to be trained in Cassini's methods of observing the eclipses of the moons of Jupiter. In 1668, Cassini had published tables listing the times of future eclipses of Jupiter's moons. The missionaries would go on to make their own observations while in Africa, the Americas, and China, and they used Cassini's tables to determine the longitudes of their remote sites. Also, other astronomers of his and later times used Cassini's methods to determine the longitudes of astronomical observation sites around the world.

Although Cassini never followed up on his own speculation that the apparent errors in the timing of the eclipses of the moons of Jupiter were a result of the finite speed of light, his measurements were used by Rømer, who announced in 1676 that he had determined the speed of light.

Cassini's achievements were so notable that the National Aeronautics and Space Administration (NASA) named its Saturn orbiter after him; the orbiter was launched in October, 1997. In addition, named in his honor are a crater on the Moon, a crater on Mars, and the Cassini Regio region on Saturn's moon Iapetus.

George J. Flynn

Further Reading

Alexander, A. F. O'Donel. *The Planet Saturn: A History of Observation, Theory, and Discovery*. New York: Dover, 1980. This work describes Cassini's many contributions to the understanding of Saturn, its rings, and his discovery of the moons Iapetus, Rhea, Dione, and Thetys.

Beatty, J. K. "A Comet Crash in 1690?" *Sky and Telescope* 93 (April, 1997): 111. Beatty summarizes a new Japanese interpretation of Cassini's 1690 sketches of Jupiter, suggesting Cassini saw the results of a comet impact.

Débarbat, S., and C. Wilson. "The Galilean Satellites of Jupiter from Galileo to Cassini, Rømer, and Bradley." In *Planetary Astronomy from the Renaissance to the Rise of Astrophysics*, edited by René Taton and Curtis Wilson. New York: Cambridge University Press, 1989. Part A of this edited volume provides an excellent account of Cassini's contributions to the measurement of the positions and eclipses of the moons of Jupiter.

Schorn, Ronald A. *Planetary Astronomy: From Ancient Times to the Third Millennium*. College Station: Texas A&M University Press, 1999. This work describes Cassini's numerous contributions to the observations of the planets, particularly Jupiter and Saturn.

NICOLAUS COPERNICUS

Polish astronomer

Copernicus dismissed the Ptolemaic system and introduced the theory that the planets, including Earth, revolve around the sun. He defended the rights of the educated to discuss scientific theories, even when those theories differ from currently accepted beliefs and contradict religious dogma.

Born: February 19, 1473; Thorn, Prussia (now Toruî, Poland)
Died: May 24, 1543; Frauenburg, East Prussia (now Frombork, Poland)
Also known as: Mikołaj Kopernik
Areas of achievement: Astronomy, church reform

Early Life

The family origins of Nicolaus Copernicus (nihk-uh-LAY-uhs kuh-PUHR-nih-kuhs) and the commercial interests of his hometown, Thorn, reflect the dual claim that Germans and Poles alike have on him. His father, also named Mikołaj (Nicolaus) Kopernik, was an immigrant from Kraków who married a daughter of a prominent burgher family, Barbara Watzenrode, and like other Thorn merchants, prospered from the exchange of Hanseatic goods for the wheat, cattle, and other produce of Poland. Thorn burghers were subjects of the Polish king, but Polish tradition allowed associated lands such as Prussia to govern themselves autonomously. Consequently, they made their political wishes felt through their representatives in the Prussian diet rather than directly to the king.

Had Mikołaj not died in 1483, his sons, Andreas and Nicolaus, would probably have entered into careers in commerce. Their guardianship, however, fell to their uncle, Bishop Lucas Watzenrode of Ermland (Warmia), who was best able to provide for them a future in church administration. A university education being indispensable to holding church offices, Bishop Lucas sent the boys to study first in Kraków, then in Italy. Nicolaus not only became a master of mathematics and astronomy but also acquired knowledge of medicine, painting, and Greek.

On his return to Prussia in 1503, Nicolaus followed the contemporary practice of Latinizing his name, Copernicus, and became one of the canons in the Ermland cathedral chapter. As his uncle's physician, assistant, and heir apparent, Copernicus was present during inspection tours, provincial diets, and royal audiences. For several years, he managed the diocese efficiently but without enthusiasm—his uncle was a hard taskmaster who lacked a sense of humor. Eventually, Copernicus announced that his interests in astronomy were greater than his ambition to become a bishop. From that time on, like most of the other canons, he lived according to clerical rules but remained a simple administrator who had no thought of becoming a priest.

Life's Work

The first of several portraits made during his lifetime show Copernicus to have been handsome and dressed in simple but elegant clothing, with nothing of either the cleric or the dandy about him. He was so utterly unremarkable in other respects that few anecdotes about him exist, leaving relatively little information about his personal life and intellectual development. Yet two facts stand out. First, Copernicus was a Humanist whose

Nicolaus Copernicus. (Library of Congress)

closest friends and associates were poets and polemicists. His translation of an ancient author, Theophilactus Symocatta, from Greek into Latin was the first such publication in the Kingdom of Poland, and he dedicated the work to his Humanistically trained uncle, Bishop Lucas. Copernicus later used Humanist arguments to defend his astronomical theories.

Second, Copernicus must be seen as a bureaucrat whose busy life made it difficult for him to make the observations of the heavens on which his mathematical calculations were based. At one time or another, he was a medical doctor, an astrologer, a cartographer, an administrator of episcopal lands, a diplomat, a garrison commander in wartime, an economic theorist, an adviser to the Prussian diet, and a guardian to numerous nieces and nephews.

About 1507, Copernicus seems to have become persuaded that the Ptolemaic system (which asserted that Earth was the center of the universe) was incorrect. From that point on, he spent every spare moment trying to demonstrate the correctness of his insight that the sun was the center of the planetary movements (the solar system).

His first description of his theory, the *Commentariolus* (1514; English translation, 1939), circulated among his friends for many years. Eventually, it came to the ears of Cardinal Schönberg, who wrote a letter asking Copernicus to publish a fuller account. This letter was ultimately published in *De revolutionibus orbium coelestium* (1543; *On the Revolutions of the Heavenly Spheres*, 1952; better known as *De revolutionibus*) as a proof that high officials in the papal curia approved of scholars discussing the existence of a solar system. Copernicus made no answer. Instead, he asked his bishop to assign him light duties at some parish center, where he could make his observations and concentrate on mathematical calculations. This request was difficult to grant because Copernicus was known to be one of the more capable diocesan administrators.

For several years, his work was interrupted by war. In 1520, the last grandmaster of the Teutonic Order, Albrecht of Brandenburg, made a final effort to reestablish his religious order as ruler of all Prussia. Copernicus led the defense of Allenstein (Olsztyn) and participated in the peace negotiations. In 1525, Albrecht, defeated at every turn, secularized the Teutonic Order in Prussia and became a Protestant vassal of the king of Poland. This brought about an immediate improvement of Albrecht's relationship to the rest of Prussia. Albrecht later called on Copernicus's services as physician, and in 1551, he published a volume of Copernicus's astrological observations.

Copernicus labored for several years to restore order to the war-ravaged Ermland finances. He advised the Prussian diet to reform the monetary system, explaining that since everyone was hoarding good coins and paying taxes with debased coins, the income of the diet was being reduced significantly. Having expounded this early version of English financier Sir Thomas Gresham's law, he recommended that all coins be called in and new ones issued. The diet, aware that it did not have the bullion to mint a sufficient number of full-weight coins, took no action. There were other, more pressing problems: politics and religion.

The spread of Lutheran reforms through Poland was halted by royal action, but not before many cities and some prominent nobles had become Protestant. The ensuing era was filled with strident debate as fanatics on both sides denounced their opponents and demanded that all parties commit themselves to what they perceived as a struggle against ultimate evil. Copernicus sought to avoid this controversy but could not. When Ermland bishop Johann Dantiscus sought to rid himself of all canons who gave any appearance of Protestant leanings, his eye fell on Copernicus, whose friends were corresponding with prominent Protestants and who, moreover, had as his housekeeper a young woman with children. Copernicus responded that his housekeeper was a widowed relative who could have no interest in a man as aged as he, but he argued in vain. He dismissed his housekeeper and watched as his friends went into exile. His health failing, Copernicus was indeed isolated from friends and family.

In 1539, a Lutheran mathematician at Wittenberg, Rheticus, made a special journey to Frauenburg to visit Copernicus. Finding him ill and without prospect of publishing the manuscript he had completed at great labor, Rheticus extended his stay to three months so that he could personally copy the manuscript. He then arranged for the publication of *Narratio prima de libris revolutionum* (1540; *The First Account*, 1939) in Danzig and for

Copernicus's Heliocentric Solar System

Nicolaus Copernicus moved against the tide of his time by proposing in his De revolutionibus, *known in English as* On the Revolutions of the Heavenly Spheres, *that the sun was at the center of the known universe, a theory that countered the longstanding geocentric theory of Ptolemy and of Scripture.*

Why should we hesitate to grant it [Earth] a motion, natural and corresponding to its form, rather than assume that the whole world [the universe], whose boundary is not known and cannot be known, moves? And why are we not willing to acknowledge that the appearance of a daily revolution belongs to the heavens, its actuality to the earth. . . .

In the middle of all dwells the Sun. Who indeed in this most beautiful temple would place the torch in any other or better place than one whence it can illuminate the whole at the same time? Not ineptly, some call it the lamp of the universe, others its mind, others again its ruler—Trismegistus, the visible God, Sophocles' Electra the contemplation of all things. And thus rightly inasmuch as the Sun, sitting on a royal throne, governs the circumambient family of stars. . . . We find, therefore, under this orderly arrangement, a wonderful symmetry in the universe, and a definite relation of harmony in the motion and magnitude of the orbs, of a kind it is not possible to obtain in any other way.

Source: The Portable Renaissance Reader, edited by James Bruce Ross and Mary Martin McLaughlin (New York: Viking Press, 1968), pp. 592-593.

the publication of the mathematical section in Nuremberg in 1542. Unable to supervise the printing of the theoretical section personally, Rheticus gave that task to another Protestant scholar, Andreas Osiander of Wittenberg.

Osiander was at a loss as to how to proceed. He saw that Copernicus had not been able to prove his case mathematically. Indeed, it would have been difficult for him to do so without inventing calculus (which was later created by Gottfried Wilhelm Leibniz and Sir Isaac Newton independently of each other for the very purpose of calculating the elliptical orbits of the planets). Consequently, Copernicus had defended his ideas by demonstrating that Ptolemy's was not the only ancient theory describing the universe; indeed, there were ancient philosophers who believed that the sun was the center of the solar system. Moreover, he had argued that free inquiry into science was as necessary as freedom to write literature or produce fine art. In this respect, Copernicus was presenting his case to Renaissance Humanists, especially to the well-educated pope to whom he dedicated his book, as a test of free thought.

Osiander, who perceived that the Catholic world was hostile to all innovations and was equally well aware of the debates raging in the Protestant world over biblical inerrancy, saw that Copernicus was treading on dangerous ground by suggesting an alternate view of the universe than the one presented in Scripture. Consequently, there was a real danger that the theory would be rejected entirely without having been read. To minimize that possibility, he wrote an unauthorized introduction that readers assumed was by Copernicus. This stated that the solar system was merely a hypothesis, a way of seeing the universe that avoided some of the problems of the Ptolemaic system. This led to much confusion and angered Copernicus considerably when he saw the page proofs. Copernicus, however, was too weak and ill to do anything about it. With a justice that is all too rare in this world, a copy of *De revolutionibus* arrived in time for him to know that his life's work was to survive.

Significance

Copernicus's theory was not immediately accepted, and not because of the controversies of the Reformation alone—although they made it dangerous for any scientist to suggest that the biblical descriptions of the heavens were incorrect. Copernicus's idealistic belief that God would create only perfectly circular planetary orbits made it impossible for him to prove his assertions mathematically. Nevertheless, Copernicus's theory was the only one to offer astronomers a way out of a Ptolemaic system of interlocking rings, which was becoming impossibly complex. His insights undermined the intellectual pretensions of astrology and set astronomy on a firm foundation of observation and mathematics.

Although Copernicus's defense of the freedom of inquiry was less important in the struggle against religious dogmatism than later demonstrations of the existence of the solar system, Copernicus became a symbol of the isolated and despised scientist who triumphs over all efforts by religious fundamentalists to silence him.

William Urban

Further Reading

Armitage, Angus. *Sun, Stand Thou Still: The Life and Works of Copernicus, the Astronomer*. New York: Henry Schuman, 1947. The best-known of many biographies, its explanation of the conceptual problems facing Copernicus is easily followed by any good reader.

Barrett, Peter. *Science and Theology Since Copernicus: The Search for Understanding*. Reprint. Poole, Dorset, England: T&T Clark, 2003. Traces the legacy of Copernicus over four hundred years. Examines the history of the debate between science and Christianity, attempting to fashion a philosophical basis for the simultaneous embrace of scientific method and religious faith in the modern world.

Beer, Arthur. *Copernicus Yesterday and Today*. Elmsford, N.Y.: Pergamon Press, 1975. A collection of useful essays that were delivered during the Copernicus celebration.

Copernicus, Nicolaus. *Three Copernican Treatises: The Commentariolus of Copernicus, The Letter Against Werner, The Narratio Prima of Rheticus*. Edited and translated by Edward Rosen. 3d ed., rev. New York: Octagon Books, 1971. This timeless translation of basic documents relating to Copernicus's achievement is accompanied by an extensive learned commentary. Rosen demonstrates that Copernicus put forward a "hypothesis" rather than a "theory" out of a fear of arousing opposition from religious fundamentalists rather than from any doubt that he was right. The third edition includes a biography of Copernicus and bibliographies of works written about Copernicus between 1939 and 1970.

Gingerich, Owen. *The Book Nobody Read: Chasing the Revolutions of Nicolaus Copernicus*. New York: Walker, 2004. A fascinating and original work of scholarship. Gingerich spent years tracking down and examining every extant copy of the original printing of Copernicus's *De revolutionibus*. Using this bibliographic analysis, he demonstrates who read the work, what they thought of it, and how exactly Copernicus's ideas spread throughout Europe. Includes illustrations, photographic plates, maps, bibliographic references, and index.

Henry, John. *Moving Heaven and Earth: Copernicus and the Solar System*. Cambridge, England: Icon, 2001. Argues that Copernicus's discovery had revolutionary effects for the cultural status afforded to theoretical science and mathematics in Western culture. He asserts that before Copernicus, pure knowledge was believed to come only from the traditions of ancient scholars, whose work was preserved only in fragments. Copernicus demonstrated that abstract mathematics and formal scientific inquiry could produce pure knowledge on their own, thereby transforming the nature of thought and truth in the West. Includes illustrations and bibliographic index.

Kesten, Hermann. *Copernicus and His World*. New York: Roy, 1945. This biography deals with Copernicus's contemporaries as much as with the astronomer himself. Kesten presents Copernicus as a warrior in the contest between science and religion. He concludes with chapters on Giordano Bruno, Tycho Brahe, Johannes Kepler, and Galileo.

Rusinek, Michat. *The Land of Nicholas Copernicus*. Translated by A. T. Jordan. New York: Twayne, 1973. The text is relatively sparse, but the pictures are unequaled in quality. The author traces the life of the as tronomer through photographs of cities, castles, and personal possessions.

Stachiewicz, Wanda M. *Copernicus and the Changing World*. New York: Polish Institute, 1973. The four hundredth anniversary of Copernicus's birth brought forth many publications. This one is unique.

ANDRÉ-LOUIS DANJON

French astronomer

Danjon primarily worked to increase the precision of astronomical instruments in observing stars. The invention of the prismatic 60-degree astrolabe, or the Danjon astrolabe, is his crowning achievement. Toward the end of his career, in the 1950's, he also investigated irregularities in Earth's rotation.

Born: April 6, 1890; Caen, France
Died: April 21, 1967; Paris, France
Areas of achievement: Astronomy, invention and technology

Early Life

André-Louis Danjon (ahn-dray-lwee dahn-zhon) attended the École Normale Supérieure in Paris from 1910 until the outbreak of World War I in 1914. During the war, he served in the French army. After his release from the military in 1919, he became an astronomer at the University Observatory at Strasbourg and continued his studies. In 1928, he obtained his doctorate degree there and started a rather rapid rise through the ranks of his profession. In 1930, he accepted the position of director of the University Observatory at Strasbourg, and a year later the University of Strasbourg offered him a position on the faculty. By 1935, Danjon occupied the dean's chair of the faculty of sciences at the university, only seven years after being granted his degree.

Life's Work

Visual photometry interested Danjon from the beginning of his studies in astronomy. This led him to various successful attempts to improve current observational hardware, producing a greater precision in his readings. In 1921, he built a photometer for observations of the lunar eclipse. Other photometer innovations allowed Danjon to be among the first to calculate Earth's reflectivity of the sun's light, or albedo, and to establish the corresponding constants for the planets Mercury and Venus.

After World War II, in 1945, Danjon left the University of Strasbourg for the directorship of the Paris Observatory. There he brought his talent for the technical improvement of instrumentation and his interests in positional astronomy together. Danjon concluded that the instrument then used for the determination of the coordinates of fundamental stars, the transit instrument, had reached its maximum precision. He therefore set out to create a new system that eliminated previous shortcomings by relying on completely different principles. His innovation, the prismatic 60-degree astrolabe, did not rely on the passage of one heavenly body across another, as before. Instead, Danjon's instrument used a prism that created two images of the same star on a graph, one descending and one ascending as Earth rotated. When these images aligned horizontally, the time at which the star attained the altitude of 60 degrees in Earth's sky could be calculated with great accuracy. Danjon utilized a fixed altitude to minimize the error caused by variations from standard atmospheric refraction. The final version of the Danjon astrolabe, as it came to be known, proved itself so worthy that four years after its introduction at the Paris Observatory in 1956 more than thirty major observatories around the world had switched to its use.

The ability of this new device to determine more precisely star positions opened new fields of interest for Danjon, for it revealed irregularities in Earth's rotation. The rate of Earth's rotation can be measured using universal time, which utilizes positional astronomy (of which Danjon's astrolabe is a part), or atomic time, which relies on atomic decay, which is constant. On occasion, scientists could determine slight differences between these two measuring techniques. In a moment of intuition, Danjon deduced that the variance of universal time from atomic time resulted from an increase in the sun's energy. In 1958, Danjon published his results, which concluded that Earth suddenly increased its rotation in coincidence with intense solar activity on February 23, 1956.

Significance

Danjon, throughout his career, split his time between advancing the state of astronomy and supporting the efforts of other astronomers. Following his own interests, his invention of the prismatic 60-degree astrolabe allowed the compilation of the coordinates of fundamental stars and facilitated the study of Earth's movement around its center of gravity. In this regard, his achievements should not be counted as slight merely because they seem few in number. To determine the reflectivity of Mercury and Venus, for example, required more than thirty thousand observations spread over ten years' time, sufficient testimony to a scientist's tenacity and productivity.

In addition to his own work, however, Danjon unselfishly spent much of his time encouraging the work of his fellows and students in various faculty and administrative positions. As an example of his directing abilities, he increased the staff of the Paris Observatory from 60 scientists to 350 in eighteen years. In addition to the time already mentioned that he spent at the observatories and universities at Strasbourg and Paris, Danjon taught at the Sorbonne and was in 1954 appointed the director of the Astrophysics Institute in Paris. His countrymen's pride in his contributions reveals itself in his election to the French Académie des Sciences in 1948, and he was the president of the International Astronomical Union from 1955 to 1958. Danjon retired from scholarly pursuits in 1963 and died four years later.

James Owen

Further Reading

Flammarion, Gabrielle Camille, and André Danjon, eds. *The Flammarion Book of Astronomy.* Translated by Annabel Pagel and Bernard Pagel. New York: Simon & Schuster, 1964. Though a general work on astronomy, it focuses on the solar system. Danjon worked on the chapters that cover Earth, the moon, and the planets. Contains many photographs and illustrations on all aspects of astronomy, including those from popular accounts throughout the ages.

Krisciunas, Kevin. *Astronomical Centers of the World.* New York: Cambridge University Press, 1988. Taking a historical look at some of the major observatories of the world, this book includes a chapter on the Paris Observatory, where Danjon worked. Its coverage of modern installations includes photographs of the telescopes used in each.

Kuiper, Gerard P., and Barbara M. Middlehurst, eds. *Telescopes.* Chicago: University of Chicago Press, 1960. Danjon wrote the chapter "The Impersonal Astrolabe" that details the design and effectiveness of the instrument for which he is famous, the prismatic 60-degree astrolabe. He also discusses its advantages over the transit instrument that it replaced. Includes photographs and a fairly detailed blueprint of the device, enumerating its numerous components. Bibliography.

Ronan, Colin A. *Astronomy.* New York: Barnes & Noble Books, 1973. A unique scrapbook of astronomical history through the ages. Extracts paragraphs and illustrations from the works of the major astronomers of the world, allowing readers to follow the thoughts of such figures as Ptolemy, Nicolaus Copernicus, Johannes Kepler, and Galileo, among others, exactly as they wrote them. Each excerpt is annotated and footnoted. Completed by superb illustrations, this chance to experience astronomy as astronomers see it should not be missed.

Sagan, Carl. *Cosmos.* New York: Random House, 1980. Sagan has ably attempted to paint the universe, and Earth's and humankind's place in it. Certainly a far-reaching book, marked by the author's excellent text. Splendid photographs and paintings. Astronomy, its history, and its significance to life on Earth are presented here in an effective and powerful manner.

Sinnot, Roger W. "Seeking Thin Crescent Moons." *Sky and Telescope* 107, no. 2 (February, 2004): 102-106. Includes information about Danjon's astronomical discoveries.

Struve, Otto, and Velta Zebergs. *Astronomy of the Twentieth Century*. New York: Macmillan, 1962. Explores the advances made in astronomy since roughly 1900 from the perspective of some of Danjon's contemporaries, astronomers who, like him, participated in these advances. Taking a historical approach, in the form of storytelling and using the scientists' own words, it presents a unique view of how scientific inquiry proceeds by a curious combination of chance discovery, fortuitous accident, and hard work.

ALBERT EINSTEIN

German-born American physicist

Einstein was the principal founder of modern theoretical physics, and his theory of relativity fundamentally changed the understanding of the physical world. His stature as a scientist, together with his strong humanitarian stance on major political and social issues, including nuclear energy and weaponry, made him one of the outstanding thinkers of the twentieth century.

Born: March 14, 1879; Ulm, Germany
Died: April 18, 1955; Princeton, New Jersey
Areas of achievement: Physics, astronomy, mathematics, peace advocacy

Early Life

Albert Einstein (IN-stin) was born in Ulm, Germany, to moderately prosperous Jewish parents. His early childhood did nothing to suggest future greatness; he was late learning to speak, and his parents feared that he might be backward. He was apparently fascinated at the age of five by the mysterious workings of a pocket compass, and at the age of twelve he became enthralled by a book on Euclidean geometry. In his childhood, he also learned to play the violin and so acquired a love of music that was to last throughout his life.

In 1888, Einstein was sent to the Luitpold gymnasium in Munich, but he disliked the regimented and authoritarian atmosphere of the school. Even at this young age he seems to have exhibited the independence of mind, the ability to question basic assumptions and to trust in his own intuition, which were to lead him to his brilliant achievements. He left the gymnasium in 1895, without gaining a diploma. His father tried to send him to a technical school in Zurich, but he failed the entrance examination, in spite of high scores on mathematics and physics. The following year he was more successful, and in 1900, he received the diploma that qualified him to teach. To his disappointment, however, he failed to obtain a teaching position at the school.

In 1901, Einstein became a Swiss citizen. (He had renounced his German citizenship in 1896.) In 1902, after temporary positions at schools in Winterthur and Schaffhausen, he secured a post as a technical expert in the Swiss patent office in Berne, where he was to remain for seven years. In the following year, he married Mileva Maric, a friend from his student days in Zurich, and in 1904 the first of their two sons was born.

Einstein's first scientific paper had been published in 1901, and he had also submitted a Ph.D. thesis to the University of Zurich. While he was working quietly in the patent office, however, isolated from the mainstream of contemporary physics, there was little to suggest the achievements of 1905, which were to shake the scientific world to its core.

Albert Einstein. (Library of Congress)

Einstein on the Scientific Impulse

In an article published in Scientific American *in 1950—more than three decades after he first presented his general theory of relativity—Albert Einstein philosophized on the human impulse toward theoretical science.*

What, then, impels us to devise theory after theory? Why do we devise theories at all? The answer to the latter question is simply: Because we enjoy "comprehending," i.e., reducing phenomena by the process of logic to something already known or (apparently) evident. New theories are first of all necessary when we encounter new facts which cannot be "explained" by existing theories. But this motivation for setting up new theories is, so to speak, trivial, imposed from without. There is another, more subtle motive of no less importance. This is the striving toward unification and simplification of the premises of the theory as a whole. . . .

There exists a passion for comprehension, just as there exists a passion for music. That passion is rather common in children, but gets lost in most people later on. Without this passion, there would be neither mathematics nor natural science. Time and again the passion for understanding has led to the illusion that man is able to comprehend the objective world rationally, by pure thought, without any empirical foundations—in short, by metaphysics. I believe that every true theorist is a kind of tamed metaphysicist, no matter how pure a "positivist" he may fancy himself. The metaphysicist believes that the logically simple is also the real. The tamed metaphysicist believes that not all that is logically simple is embodied in experienced reality, but that the totality of all sensory experience can be "comprehended" on the basis of a conceptual system built on premises of great simplicity. The skeptic will say that this is a "miracle creed." Admittedly so, but it is a miracle creed which has been borne out to an amazing extent by the development of science.

Life's Work

In 1905, Einstein published three major papers, any one of which would have established his place in the history of science. The first, which was to bring him the Nobel Prize in Physics in 1921, explained the photoelectric effect and formed the basis for much of quantum mechanics. It also led to the development of television. The second concerned statistical mechanics and explained the phenomenon known as Brownian motion, the erratic movement of pollen grains when immersed in water. Einstein's calculations gave convincing evidence for the existence of atoms.

It was the third paper, however, containing the special theory of relativity, that was to revolutionize understandings of the nature of the physical world. The theory stated that the speed of light is the same for all observers, and is not dependent on the speed of the source of the light, or of the observer, and that the laws of nature (both the Newtonian laws of mechanics and Maxwell's equations for the electromagnetic field) remain the same for all uniformly moving systems. This theory meant that the concept of absolute space and time had to be abandoned because it did not remain valid for speeds approaching those of light. Events that happen at the same time for one observer do not do so for another observer moving at high speed in respect of the first. Einstein also demonstrated that a moving clock would appear to run slow compared with an identical clock at rest with respect to the observer, and a measuring rod would vary in length according to the velocity of the frame of reference in which it was measured.

In another paper published in 1905, Einstein stated, by the famous equation $E = mc2$, that mass and energy are equivalent. Each can be transferred into the other because mass is a form of concentrated energy. This equation suggested to others the possibility of the development of immensely powerful explosives.

Such was Einstein's achievement at the age of twenty-six. There had not been a year like it since Newton published his *Principia* in 1687. The scientific world quickly recognized him as a creative genius, and in 1909 he took up his first academic position, as associate professor of theoretical physics at the University of Zurich. After two more positions, one in Prague and the other in Zurich, he became a member of the Prussian Academy of Sciences and moved to Berlin in 1914.

In the meantime, Einstein had been working to extend the special theory of relativity to include new laws of gravitation, and the general theory of relativity was published in 1916. It was one of the greatest intellectual productions ever achieved by one person, and its picture of the universe as a four-dimensional space-time continuum lies at the foundation of all modern views of the universe. The theory stated that large masses produce a gravitational field around them, which results in the curvature of space-time. This gravitational field acts on objects and on light rays; starlight, for example, is deflected when passing through the gravitational field of the sun.

In 1919, the general theory received experimental confirmation from a team of British astronomers.

Suddenly, the world awoke to the implications of Einstein's work, and he found himself internationally celebrated as the greatest scientist of the day. During the early 1920's, he traveled extensively in Europe, the Far East, and the Americas, hailed everywhere as genius, sage, and hero. With his untidy shock of hair—formerly black, now graying—rising from a high forehead, his deep brown eyes, and small moustache, he made a striking figure. It was not only his superior intellect that aroused public recognition and respect but also his simple good nature, nobility, and kindliness. However, Einstein, always modest, was genuinely astonished at the attention he received.

It was during his travels in the 1920's that the other great concerns of Einstein's life came to the fore. A man of deep humanitarian instincts, he did not isolate himself from the turbulent political events around him. During World War I, he had spoken out against militarism and nationalism. Now, as a famous person, he once more took up the cause of pacifism, expressing his opinion openly, caring nothing for popularity. Einstein's other lasting concern was the promotion of Zionism, and his tour of the United States in 1921 was undertaken in part to raise funds for Hebrew University. These activities made him a target for fierce abuse from the Nazis, and even outside Germany, his radical political views made him a controversial figure.

When Hitler came to power in 1933, Einstein was on his third visit to the United States, and he resolved not to return to Germany. After brief stays in Belgium and England, he left Europe for the last time, to become a professor at the Institute for Advanced Study at Princeton. He continued to lend his support to the cause of justice and freedom, helping Jewish refugees whenever he could, and he modified his former pacifism in the face of the threat of Nazi domination. In 1939, he was persuaded to write a letter to President Franklin D. Roosevelt, alerting him to the military potential of atomic energy. (Einstein played no part, however, in the research that led to the development of the atomic bomb, which merely verified the truth of his famous equation.) After the war, he remained tirelessly devoted to the cause of world peace, and proposed a world government in which all countries were to agree to forfeit part of their national sovereignty.

His later scientific career took two main directions. First, he was so deeply convinced of nature's fundamental simplicity that he labored unsuccessfully for thirty years in an attempt to construct a unified field theory. Second, he could not accept one of the fundamental results of quantum theory, that the interaction of subatomic particles could be predicted only in terms of probabilities. "God does not play dice with the world," he remarked. Many of his colleagues thought him stubborn, but nevertheless he remained a revered figure; his reputation as a genius who also possessed wisdom and saintliness never left him, neither in life nor in death.

Significance

Einstein's scientific achievements place him alongside such figures as Copernicus, Galileo, and Newton, as one who vastly enlarged the scope of human knowledge about the physical universe. In this respect he is a universal figure and belongs to no country. It is perhaps appropriate, however, that Einstein, a German Jew to whom destiny had decreed a nomadic existence, eventually found a permanent home in the United States. His links with his adopted country—he became a U.S. citizen in 1941—are profound. He was the most illustrious of the hundreds of intellectuals who fled from Europe before World War II, and his presence at the newly formed Institute for Advanced Study, which marked a new period of development for American research and education, played a key role in attracting other eminent scholars.

Einstein had always viewed the United States as a bulwark of democracy and individual freedom, and in the debates that divided the country in the postwar decade—particularly debates on the Cold War and the use of nuclear energy—his was a consistent voice for sanity and decency in human affairs. He spoke out for freedom of thought and speech in the McCarthy era, when he feared that the United States was betraying its own ideals, and he continued to urge scientists to consider the social responsibilities of their work in the atomic age. In his final year, he and a group of leading scientists signed a statement, known as the Russell-Einstein Manifesto, warning about the terrible consequences of nuclear war. This led to the Pugwash Conference on science and world affairs in 1957, in which for the first time scientists from East and West met to discuss nuclear arms. A series of influential conferences followed, and the Pugwash movement has continued its activities ever since.

Bryan Aubrey

Further Reading

Bernstein, Jeremy. *Einstein.* New York: Viking Press, 1973. First published in *The New Yorker* magazine, this biography includes a historical survey of the scientific issues of Einstein's life. No index.

Brian, Denis. *The Unexpected Einstein: The Real Man Behind the Icon*. Hoboken, N.J.: J. Wiley, 2005. Based on interviews with friends and associates, this biography demystifies Einstein to present a realistic portrait of his life, thoughts, and personality.

Clark, Ronald W. *Einstein: The Life and Times*. New York: World, 1971. Lengthy, authoritative, and readable. Gives a balanced treatment of Einstein's life, although the author tends to see Einstein as an idealist who was out of his depth when he entered practical politics.

Einstein, Albert. *Out of My Later Years*. 1950. Rev. ed. Westport, Conn.: Greenwood Press, 1970. Collection of Einstein's writings, including articles, letters, appeals, and miscellaneous papers, from 1934 to 1950, on the philosophical, political, social, and scientific issues of the period.

Einstein, Albert, and Leopold Infeld. *The Evolution of Physics*. 1938. New ed. New York: Simon & Schuster, 1961. Covers the period from the rise of the mechanical philosophy to relativity and quanta. Einstein's clarity of thought and expression is apparent throughout, and this remains one of the best popular expositions of his thought. No mathematics.

French, A. P., ed. *Einstein: A Centenary Volume*. Cambridge, Mass.: Harvard University Press, 1979. Beautifully designed, lavishly illustrated. Includes personal reminiscences from Einstein's friends and colleagues; expositions of many aspects of his scientific, political, and educational thought; and extracts from his writings.

Isaacson, Walter. *Einstein: His Life and Universe*. New York: Simon & Schuster, 2007. Isaacson, the prominent biographer of Benjamin Franklin and other subjects, discusses how Einstein's love for intellectual and individual freedom and his rebellious nature formed his scientific imagination.

Robinson, Andrew, et al. *Einstein: A Hundred Years of Relativity*. New York: Harry N. Abrams, 2005. One of the better books released to commemorate the one-hundredth anniversary of Einstein's scientific paper outlining the theory of relativity. This volume includes essays by prominent physicists, including Stephen Hawking and Freeman Dyson, who discuss Einstein's theories and influence. It also features photographs of Einstein bequeathed by him to Hebrew University in Jerusalem.

Sayen, Jamie. *Einstein in America: The Scientist's Conscience in the Age of Hitler and Hiroshima*. New York: Crown, 1985. A well-researched, detailed account of Einstein's nonscientific activities from 1933 to 1955. Gives political and historical background on issues such as Palestine, the Cold War, and McCarthyism.

Schwartz, Joseph. *Einstein for Beginners*. New York: Pantheon Books, 1979. Amusing, entertaining, comic-book format, useful for those with no scientific background. Clear explanation of special relativity, and some biographical background. Illustrated by Michael McGuiness.

EUDOXUS OF CNIDUS

Greek geometer

Eudoxus and his disciples resolved classical difficulties in the fields of geometry and geometric astronomy. Their approach became definitive for later research in these fields.

Born: c. 390 B.C.E.; Cnidus, Asia Minor (now in Turkey)
Died: c. 337 B.C.E.; Cnidus, Asia Minor

Early Life

As for so many ancient figures, little is known about the life of Eudoxus (yew-DAHK-suhs) of Cnidus. If one follows the account of the ancient biographer Diogenes Laertius (c. 250 C.E.), Eudoxus first visited Athens at age twenty-three to study medicine and philosophy. He soon returned home to Cnidus, however, and from there, joining the company of the Cnidian physician Chrysippus, he moved on to Egypt, where for more than a year he studied among the priests and engaged in astronomical investigations. Later, as he traveled and lectured in the wider Aegean area (specifically, Cyzicus and the Propontis), he built up a following and thus returned to Athens a man of considerable distinction. His main subsequent activity seems to have centered on Cnidus, where he was honored as a lawgiver. His renown extended to many areas, including astronomy, geometry, medicine, geography, and philosophy.

There is disagreement over his dates. The ancient chronicler Apollodorus sets Eudoxus's prime activity in 368-365 B.C.E. In general, the prime means age forty; if that holds here, Eudoxus's birth would be set c. 408. There is reason for doubt, however, because this early a date conflicts with other biographical data. G. L. Huxley favors c. 400; G. de Santillana and others argue for c. 390.

Eudoxus is reported to have died at the age of fifty-three; the corresponding date would be 355, 347, or 337.

Life's Work

None of Eudoxus's writings survives, but fragments cited by ancient authors offer a reasonable impression of their diversity and significance. His principal efforts were in the areas of mathematics and astronomy, the former best represented in portions of Euclid's *Stoicheia* (c. 300 B.C.E.; *The Elements of Geometrie of the Most Auncient Philosopher Euclide of Megara*, 1570, commonly known as the *Elements*), the latter in astronomical discussions of the fourth-century cosmology of Aristotle and the ancient commentaries on it.

According to Archimedes (287-212 B.C.E.), Eudoxus was the first to set out a rigorous proof of the theorems that any pyramid equals one-third of the associated prism (that is, having the same height and base as the pyramid), and any cone equals one-third of the associated cylinder. Eudoxus also appears to have proved two other theorems: that circles are as the squares of their diameters and that spheres are as the cubes of their diameters. The proofs of these four theorems constitute the main part of book 17 of *Elements*, and the technique used there is likely to derive from Eudoxus.

To take the circle theorem as an example, one could imagine a regular polygon having so many sides that it seems practically indistinguishable from a circle. As two such polygons (with equally numerous sides) are proportional to the squares of their diameters, the same could be supposed for the corresponding circles. Presumably, an argument of this sort was assumed by geometers who used the circle theorem in the time before Eudoxus. In the strict sense, however, the reasoning would be invalid, for only by an infinite process can rectilinear figures eventuate in the circle.

In the Eudoxean scheme, one assumes the stated proportion to be false: If two circles are not in the ratio of the squares of their diameters, then one can construct two similar regular polygons, one inscribed in each circle, and one can make the difference between the polygon and its circle so small that the polygon is found to be simultaneously greater and less than a specified amount. Because that is clearly impossible, the stated theorem must be true. (This indirect manner for proving theorems on curved figures is often called, if somewhat misleadingly, the "method of exhaustion.")

A key move in this proof is making the polygon sufficiently close to the circle. To this end, one observes that as the number of sides is doubled, the difference between the polygon and the circle is reduced by more than half. The procedure of successively bisecting a given quantity will eventually make it less than any preassigned amount, however, as Euclid proves in *Elements*. According to Archimedes, however, it seems that Eudoxus took this bisection principle as an axiom.

The notion of proportion itself runs into a similar difficulty with the infinite. As long as quantities are related to one another in terms of whole or fractional numbers (for example, if one area is 7/5 of another area), their ratios can be specified from these same numbers (that is, the ratio of the one area to the other will be 7 to 5). Yet what if no such numbers exist? For example, it was found, a century or so before Eudoxus, that the diagonal and side of a square cannot equal a ratio of whole numbers. (In modern terms, one calls the associated number $\sqrt{2}$ "irrational"; its decimal equivalent 1.414 . . . will be nonterminating and nonrepeating.) Only by means of some form of infinite sequence can "commensurable" quantities (those whose ratio is expressible by two integers) equal the ratio of incommensurable ones. Geometers in the generation before Eudoxus had pursued the study of incommensurables with considerable interest, but Eudoxus was the first to see how the theorems on ratios could be rigorously proved when their terms were incommensurable.

It is usually supposed that Eudoxus's approach was identical to that given by Euclid in book 5 of *Elements*. Other writers, in particular Archimedes, however, knew of a different technique of proportions that seems more like what Eudoxus would have proposed. By this technique, one first establishes the stated theorem for the case of commensurable quantities. For the incommensurable case, one uses an indirect argument: If the proportionality does not hold, one can find commensurable terms whose ratio differs by less than a specified amount from the ratio of the given incommensurable terms—this is done by successively bisecting one of the givens until it is less than the difference between two others; when the commensurable case of the theorem (already proved) is applied, a certain term will be found to be simultaneously greater and less than another. Because that is impossible, the theorem must be true.

The defining notions of the proportion theory in Euclid's book 5 can be derived as a simple modification of this technique, for the role that the intermediate commensurable terms play in it is assumed by the Euclidean definition of proportion: that A:B = C:D means that for all positive integers m, n, if $m\text{A} > n\text{B}$, then also $m\text{C} > n\text{D}$;

the same holds true if = or < is substituted for >. Proofs in this Euclidean manner do not require a division into commensurable and incommensurable cases, nor do they make use of the bisection principle; in general, they are rather easier to set up than in the alternative technique. It is thus possible to see Euclid's approach as an intended refinement of the Eudoxean.

In either the Eudoxean or Euclidean form, this manner of proportion theory can be made to correspond to the modern definition of real number, as formulated by the German mathematician Julius Wilhelm Richard Dedekind. In each example, the real term (possibly irrational) is considered to separate all the rationals into those greater and those less than it.

It seems likely that Eudoxus also contributed to the study of incommensurable lines. His predecessor Theaetetus (d. c. 369) had shown that if the squares of two lines are commensurable with each other but do not have the ratio of square integers, then the lines themselves will be incommensurable with each other; further, if two such lines A, B are taken, the lines A ± B will be incommensurable with them, not only as lines but also in square (lines of this latter type were called *alogoi*, literally, "without ratio"). The further study of the *alogoi* lines, as collected in book 10 of Euclid's *Elements*, divided into twelve classes all the *alogoi* formed as the square roots of R(A ± B), where R is a unit line, and A and B are commensurable with each other in square only. Presumably, Eudoxus and his followers played a part in this investigation.

Eudoxus's efforts are rooted in a concern for logical precision in geometry, and this interest may reflect his close association with the Platonic Academy at Athens. Two anecdotes (of questionable historicity) celebrate this connection. The first explains how Eudoxus came to be involved in seeking the cube duplication, the so-called Delian problem. To allay a plague, the citizens of Delos were commanded by the oracle to double a cube-shaped altar. When their attempts failed, they sent to Plato, who directed his mathematical associates Archytas, Menaechmus, and Eudoxus to solve it. When they did so, however, Plato criticized their efforts for being too mechanical. The solutions of a dozen different ancient geometers are known, but that of Eudoxus has not been preserved. It supposedly employed "curved lines" of some sort, and reconstructions have been proposed.

In a second story, Plato is said to have posed to Eudoxus the problem of "saving the phenomena" of planetary motion on the restriction to uniform circular motion. An account of Eudoxus's scheme is transmitted by Simplicius of Cilicia (sixth century C.E.) in his commentary on Aristotle's *De caelo* (before 335 B.C.E.; *On the Heavens*, 1777). From this account a reconstruction was worked out by the Italian historian of astronomy Giovanni Virginio Schiaparelli in 1875. The Eudoxean system reproduces the apparent motion of a planet by combining the rotations of a set of homocentric spheres. The planet is set on the equator of a uniformly rotating sphere. If a second sphere is set about the first, rotating with equal speed to the first but in the opposite direction and having its axis inclined, then the planet will trace out an eight-shaped curve (which the ancients called the *hippopede*, or horse fetter), so as to complete the full double loop once for every full revolution of the spheres. One superimposes over this a third spherical rotation, corresponding to the general progress of the planet in the ecliptic, and finally over this a fourth rotation, corresponding to the daily rotation of the whole heaven. In this way, each of the five planets requires four spheres, while the sun and the moon each take three.

Schiaparelli's exposition thus revealed the ingenuity of Eudoxus's scheme for reproducing geometrically the seemingly erratic forward and backward (retrograde) motion of the planets. Nevertheless, the model proves unsuccessful in some respects: Because the planets do not vary in distance from the earth (the center of their spheres), Eudoxus cannot account for their variable brightness or for asymmetries in the shape of their retrograde paths. Even worse, the values that Eudoxus had to assign for the rotations of the spheres do not produce retrogrades for Mars or Venus, and the sun and the moon are given uniform motions, contrary to observation. Apparently, the latter two defects were recognized, for Eudoxus's follower Callippus introduced seven additional spheres (two each for sun and moon, one each for Mercury, Venus, and Mars) to make the needed corrections.

The Eudoxean-Callippean scheme is enshrined in Aristotle's *Metaphysica* (c. 335-323 B.C.E.; *Metaphysics*, 1801), in which it serves as the mathematical basis of a comprehensive picture of the entire physical cosmos. Doubtless, Eudoxus proposed his geometric model without specific commitments on physical and cosmological issues. Nevertheless, it suited well the basic Aristotelian principles—for example, that the cosmos separates into two spherical realms, the celestial and, at its center, the terrestrial, and that the natural motions of matter in the central realm (for example, earthy substances moving in straight lines toward the center of the cosmos) differ from those in the outer (where motion is circular, uniform, and eternal). Ironically, these Aristotelian principles persisted in later cosmology, even after astronomers had switched from the homocentric spheres to eccentrics, epicycles, and other geometric devices.

Eudoxus also produced works of a descriptive and empirical sort in astronomy and geography. His *Phaenomena* (fourth century B.C.E.; phenomena) and *Enoptron* (fourth century B.C.E.; mirror) recorded observations of the stars—the basis, one would suppose, of a systematic almanac of celestial events (for example, solstices and equinoxes, lunar phases, heliacal risings of stars). He adopted, as Diogenes and others report, an *oktaeteris*, or eight-year calendar cycle. As known to later authors, an *oktaeteris* is one of the cycles found to reconcile the solar year of 365.25 days with the period of the moon's phases (somewhat over 29.5), by parsing out the 2,922 days in eight years into ninety-nine lunar periods (fifty-one of thirty days and forty-eight of twenty-nine). However, it is unclear whether this was the arrangement used by Eudoxus. His geographical treatise, the *Gēs periodos* (fourth century B.C.E.; circuit of the Earth), systematically described the lands and peoples of the known world, from Asia in the east to the western Mediterranean region. A connection with his astronomical studies can be seen in the use of the ratio of longest to shortest periods of daylight for designating the latitudes of places.

Significance

However interesting Eudoxus's contributions to calendarics, geography, and philosophy may be, they are secondary to his achievement in mathematics, for he may justly be viewed as the most significant geometer in the pre-Euclidean period. He advanced the study of incommensurables, introduced a new technique for generalizing the theory of proportion, and made exact the theory of limits with his new method of "exhaustion." Remarkable for the logical precision of his proofs, Eudoxus here set the standard against which even the foremost of the later geometers, such as Euclid and Archimedes, measured their own efforts.

Eudoxus's influence on geometric astronomy is more subtle. Already, early in the third century B.C.E., astronomers had discarded his system of homocentric spheres in their pursuit of viable geometric models for the planetary motions. If the shortcomings of Eudoxus's model were evident, however, it nevertheless defined for later astronomers the essence of their task: to represent the planetary phenomena by means of uniform circular motion. Eudoxus's success thus remains implicit in the later development of astronomy, from Apollonius and Hipparchus to Ptolemy.

Wilbur R. Knorr

Further Reading

Charles, David. *Aristotle on Meaning and Essence*. Oxford, England: Clarendon Press, 2000. In this work on Aristotle, Charles is critical of his subject's analysis of Eudoxus and Callippus's astronomical conclusions.

de Santillana, G. "Eudoxus and Plato: A Study in Chronology." In *Reflections on Men and Ideas*. Cambridge, Mass.: MIT Press, 1968. A revised chronology of Eudoxus's life is argued on the basis of a detailed examination of the ancient biographical data and collateral historical evidence.

Knorr, W. R. *The Ancient Tradition of Geometric Problems*. Boston: Birkhäuser, 1986. Chapter 3 considers Eudoxus's studies of exhaustion and cube duplication, discussed in the wider context of pre-Euclidean geometry.

Neugebauer, O. *A History of Ancient Mathematical Astronomy*. New York: Springer-Verlag, 1975. All facets of Eudoxus's contributions to astronomy are covered; particularly detailed is the discussion of his planetary models. Includes an index.

Waerden, B. L. van der. *Science Awakening*. 4th ed. Princeton Junction, N.J.: Scholar's Bookshelf, 1988. The author provides a brief, insightful review of Eudoxus's mathematical work.

DAVID AND JOHANNES FABRICIUS

German astronomers

David Fabricius discovered the first periodic variable star, and his son, Johannes, was the first to make telescopic observations of sunspots, although some contend it was Galileo who did so before Johannes. Johannes was, however, the first to publish on the subject of sunspots. Also, David and Johannes determined that the Sun rotates on its axis.

David Fabricius

Born: March 9, 1564; Esens, East Frisia (now in Germany)
Died: May 7, 1617; Osteel, East Frisia (now in Germany)
Also known as: David Faber

Johannes Fabricius

Born: January 8, 1587; Resterhafe, East Frisia (now in Germany)
Died: March 19, 1616; Osteel, East Frisia (now in Germany)
Also known as: Johann Goldsmid
Areas of achievement: Astronomy, science and technology, religion and theology, scholarship, geography

Early Lives

Not much is known about the early lives of David and Johannes Fabricius (fah-BREET-see-uhs). "Fabricius" is a Latin form of the surname "Faber." In the early 1580's, David Fabricius studied in Resterhafe to become a Lutheran pastor. By 1585, he was serving as a pastor in the small town of Osteel. He was the father of seven children, four daughters and three sons.

In addition to being a respected clergyman, David was an adept cartographer and compiled the first map of East Frisia in 1589. He was very interested in science, particularly astronomy, and through correspondence and personal visits became good friends with astronomers Tycho Brahe and Johannes Kepler. He met with Brahe to discuss astronomy and astrology at Wandsburg Castle (in southwestern Germany) in 1598 and also in Prague in 1601.

Johannes Fabricius, David's second child and first son, studied medicine between 1604 and 1610 in Helmstedt and Wittenberg, then in Leyden. While at Leyden, his interest in astronomy developed. He became very interested in using a telescope to make observations of stars and planets. In 1610, he returned to Osteel and began to practice as a physician, taking with him a telescope. He apparently continued some further medical studies at Wittenberg in 1611.

Lives' Work

By the mid-1590's, David Fabricius, by this time an accomplished pastor and mapmaker, also was an astute observational astronomer. On August 3, 1596, he discovered without a telescope a star originally named Omicron Ceti. The star was later renamed Mira, meaning "wonderful." Because of the star's brightness, David thought initially that Mira was a nova. With subsequent observations, he noticed that the star was sometimes bright enough to be easily seen with the naked eye, while at other times it faded out of sight. He had discovered the first periodic, pulsating variable star, a discovery that refuted Aristotle's notion that the heavens were constant and perfect.

From an observatory that he established in Osteel, David observed the supernova of 1604 near the conjunction of Jupiter and Mars and reported his findings in both German and Latin publications. Kepler referred to David in some of his writings as the second best astronomer in Europe next to Tycho Brahe.

As a practicing physician in Osteel, Johannes Fabricius devoted much of his spare time to astronomical observations using the telescope that he had obtained in Leyden. On March 9, 1611, he made the first telescopic observations of sunspots. Apparently, his father, David, had observed sunspots many years earlier without a telescope, as had Korean and Chinese astronomers and early English astronomer John of Worcester. Shortly after his first observations of sunspots, Johannes made geometrical drawings of his findings and shared them with his father. Subsequently, they made numerous additional observations, typically just after sunrise and just before sunset. Finding moving blemishes on the Sun was another refutation of Aristotle's premise that the heavens were perfect and constant.

Because of the eye inflammation and pain that came from viewing the Sun directly, David and Johannes soon adopted and perfected the camera obscura technique. With this method, an image of the Sun is projected onto a suitable surface in a darkened room by passing sunlight through a pinhole opening that is made in the covering of a window in the room. The camera obscura was a safe way to observe the sun, and, also, it gave a clearer view of the solar disk. Between March and June of 1611, David and Johannes recorded numerous observations of sunspots. They noted that spots appearing on the western edge of the Sun would reappear on the eastern side after approximately twelve days. They made the correct interpretation of sunspot movement by concluding that the Sun must rotate on its axis. The two differed in their opinion as to where the spots were located, however. Johannes believed the spots to be on the surface of the sun, whereas his father thought that they might be clouds or planetary objects passing by the Sun and not part of the Sun itself.

Johannes wrote *De maculis in sole observatis* (narration on spots observed on the sun) and published the booklet in 1611 using the name "Johann Goldsmid," a version of his name evidently used while he was studying in Leyden. Unfortunately, observation times and dates, as well as sketches of the spots, were not included in the treatise. Although it was sold at a book fair in Frankfurt in the fall of 1611, it remained in obscurity until it was rediscovered and publicized in 1723. In the meantime, Galileo and Christoph Scheiner had each claimed credit for making the first telescopic observations of sunspots during the latter part of 1611 or early 1612. Due credit was not given to David and Johannes Fabricius until more than one hundred years after their important work.

Johannes died of unknown causes in 1616 at the very young age of twenty-nine. Almost one year later, David suffered a surprising death at the hands of an irate member of his Lutheran congregation. While presenting a sermon, the elder Fabricius had chastised farmer Frerik Hoyer for stealing a goose. Later, Hoyer confronted David in the church yard and struck him in the head with a shovel, killing the pastor. David was depicted in a novel by nineteenth century French writer Jules Verne. In Verne's *Autour de la lune* (1870; *From the Earth to the Moon . . . and a Trip Around It*, 1873), David is portrayed as an astronomer who viewed inhabitants on the moon with his telescope.

In honor of the contributions of David and Johannes to astronomy, a large impact crater on the Moon was named "Fabricius." The crater is located within the northeastern part of the Janssen walled plain in the southern hemisphere of the Moon and has a diameter of approximately 90 kilometers. It is not known, however, if the crater was named originally to honor both David and Johannes Fabricius, or to honor just one.

David and Johannes Fabricius Look at the Sun

David and Johannes Fabricius looked directly at the morning and late afternoon sun through a telescope to see sunspots. Even through a telescope, the sun's rays were powerful, causing their work to be painful and debilitating. They describe this experience here, in a translation published in 1916 in Popular Astronomy.

Having adjusted the telescope, we allowed the sun's rays to enter it, at first from the edge only, gradually approaching the center, until our eyes were accustomed to the force of the rays and we could observe the whole body of the sun. We then saw more distinctly and surely the things I have described [sunspots]. Meanwhile clouds interfered, and also the sun hastening to the meridian destroyed our hopes of longer observations; for indeed it was to be feared that an indiscreet examination of a lower sun would cause great injury to the eyes, for even the weaker rays of the setting or rising sun often inflame the eye with a strange redness, which may last for two days, not without affecting the appearance of objects.

Source: W. M. Mitchell, "The History of the Discovery of Solar Spots." *Popular Astronomy* 24 (1916). Cited on the Web site of the National Center for Atmospheric Research, High Altitude Observatory, "Johann Fabricius (1587-1616)." http://www.hao.ucar.edu/public/education/sp/images/fabricius.html. Accessed December, 2004.

Significance

Astronomical observations made by David Fabricius initiated the study of periodic variable stars by other astronomers, leading to an explanation of pulsating star phenomena in terms of varying luminosity and surface temperature. Observations of sunspot phenomena made by Johannes and David Fabricius advanced the use of the camera obscura method for investigating the Sun and motivated other astronomers to study sunspots. Interpretation of the day-by-day movement of sunspots by David and Johannes Fabricius led them to conclude correctly that the Sun rotates on its axis. The younger Fabricius was the first astronomer to publish the results of such findings. Their discovery of pulsating variable stars and telescopic observations of sunspots led many astronomers to revise their own theories that the heavens were perfect and constant, as originally thought by Aristotle.

The work of David and Johannes Fabricius was noted by many contemporary astronomers, including Johannes Kepler, Michael Mästlin, and Simon Marius. David and Johannes Fabricius were instrumental in stimulating the curiosity of other astronomers to carefully study and correctly interpret observations of the stars.

Alvin K. Benson

Further Reading

Brody, Judit. *The Enigma of Sunspots: A Story of Discovery and Scientific Revolution.* Edinburgh: Floris Books, 2002. Brody relates the fascinating story of the discovery of sunspots by David and Johannes Fabricius and the efforts that have been made since their time to describe and understand sunspots. Includes bibliographical references and color illustrations.

Christianson, John Robert. *On Tycho's Island: Tycho Brahe, Science, and Culture in the Sixteenth Century.* New York: Cambridge University Press, 2002. This work tracks the history of Tycho Brahe—philosopher, chemist, and astronomer—and discusses his interaction with David Fabricius. David corresponded and visited with Brahe and discussed some of his astronomical discoveries with him. Includes bibliographical references and index.

Drake, Stillman, Noel M. Swerdlow, and Trevor Harvey Levere. *Essays on Galileo and the History and Philosophy of Science*. Vol. 1. Toronto: University of Toronto Press, 1999. This volume includes a discussion on the controversy over who discovered sunspots—David and Johannes Fabricius, Galileo, Christoph Scheiner, or someone else. Includes bibliographical references and index.

Kepler, Johannes, and Edward Rosen. *Kepler's Somnium: The Dream: Or, Posthumous Work on Lunar Astronomy*. New York: Dover Books, 2003. Recounts some of Kepler's accomplishments and ideas, including his dream about a trip to the Moon, and tells of the friendship of David and Johannes Fabricius with Kepler.

WILLIAMINA PATON STEVENS FLEMING

Scottish-born American astronomer

Fleming pioneered the analysis of stellar spectra and discovered more than three hundred variable stars and ten exploding stars (novas). Her work made her the leading woman astronomer of her time.

Born: May 15, 1857; Dundee, Scotland
Died: May 21, 1911; Boston, Massachusetts
Also known as: Williamina Paton Stevens (birth name); Williamina Fleming; Mina Fleming
Area of achievement: Astronomy

Early Life

Williamina Paton Stevens Fleming was born in Scotland to Robert Stevens and Mary Walker Stevens. Her father was a craftsperson in the carving and gilding trade and was among the first in Dundee, Scotland, to experiment with the new daguerreotype photographic process. He died when Williamina was seven.

Williamina became a pupil-teacher at the age of fourteen. In 1877, after five years of teaching, she married James Orr Fleming; the following year they emigrated to the United States and settled in Boston. After her husband abandoned her two years later, she took a job as a maid in the home of Edward C. Pickering, who was director of the Harvard College Observatory from 1877 to 1919. A few months later she took a short leave to return to Scotland for the birth of her son, Edward Pickering Fleming, who would later graduate from the Massachusetts Institute of Technology and become a mining engineer and metallurgist.

At the Harvard College Observatory, Pickering pioneered in employing women as "computers"—people who measured and analyzed the vast number of stellar-spectrum photographs he was producing. At that time, women were not allowed to work in the observatory at night, when observations of the skies were made, but they could be employed during the daytime to analyze photographs at salaries that were lower than those of male astronomers. When Pickering became impatient with the inefficiency of a male assistant, he claimed that his Scottish maid could do a better job. In 1881, he invited Fleming to work as a computer at Harvard, even though she had neither a college education nor training in astronomy. However, her work was so good that Pickering decided to hire other young women to work with her. Over the next thirty years, Fleming worked closely with Pickering, first in clerical and computing work, but advancing rapidly to become one of the leading astronomers of her day and to supervise many other women computers at Harvard.

Life's Work

During the nineteenth century astronomy underwent a major revolution. New techniques for lens making, more advanced telescope designs, and the use of photography led to measurements of fainter stars with greater precision than ever before. In 1872, the American physician Henry Draper made the first successful photograph of the spectrum of a star. After Draper died in 1882, his widow established the Henry Draper Memorial to assist the program of stellar spectroscopy at the Harvard College Observatory. When a thin prism was placed in front of the objective lens of a refracting telescope, each star formed its own spectrum on a photographic plate at the focus, making it possible to record long exposures of spectra of as many as two hundred stars on a single photographic plate. In 1886, Fleming was placed in charge of the Draper project to measure, analyze, and classify stellar spectra.

Fleming's chief assignment was classifying stars on the basis of the various patterns of lines and bands that appeared in their spectral photographs. Because photographs were more sensitive than the human eye, they revealed new features that required a more complex classification scheme than earlier systems based solely on visual observations. In the "Pickering-Fleming system" light spectra were organized into fifteen categories, using most of the letters from A to Q according to the complexity of their lines and bands, but about 99 percent of the stars were in the six classes designated A, B, F, G, K, and M.

In 1890, the *Draper Catalogue of Stellar Spectra* was published as volume 27 of the *Annals of Harvard College Observatory*. Fleming's earlier work had been published in the customary way under the name of the director of the observatory. However, by the year 1890, the signature "M. Fleming" began to appear along with Pickering's on reports in astronomy journals. In the *Draper Catalogue*, Pickering acknowledged that "Mrs. M. Fleming" did most of the work in measuring and classifying spectra and preparing the catalog for publication. In the preparation of the catalog, Fleming used 633 photographic plates taken with the eight-inch Bache telescope. She measured 28,266 spectra and classified 10,351 stars brighter than seventh magnitude.

In the course of her work on the *Draper Catalogue*, Fleming found that the spectra of many variable red stars contained bright emission lines in addition to the usual dark absorption lines. Her further investigation showed that all stars with such unusual spectral features are variable in their luminosity. Her finding provided a new tool for discovering long-period variables. Using this technique, she discovered 222 variable stars. In her 1907 paper "A Photographic Study of Variable Stars," which was published as volume 47 of the *Annals of Harvard College Observatory*, she gave the variation in brightness of each star based on a selected list of companion stars.

In addition to Fleming's work on variable stars, she made a special study of other stars with peculiar spectra. This led her to the discovery of ten novas—or exploding stars—more than a third of the twenty-eight novas known at the time of her death. She also discovered 94 of the 107 known Wolf-Rayet stars, which are massive stars characterized by broad emission lines. Shortly after she died, her list of "Stars Having Peculiar Spectra" was published as volume 56 of the *Annals of Harvard College Observatory* in 1912.

Fleming combined her diligence and research ability with effective administrative skill in her supervision of several dozen young women. She also served as editor of all the publications issued by the Harvard College Observatory. In 1898, the Harvard Corporation named her curator of astronomical photographs. Under her supervision, nearly 200,000 photographic plates taken at Cambridge and in Peru between 1882 and 1910 were examined and cataloged. In 1906, she became the first American woman elected to the British Royal Astronomical Society, and she was also a charter member of the American Astronomical and Astrophysical Society. In 1911 she became ill with pneumonia and died in Boston at the age of fifty-four.

Significance

Williamina Fleming's discovery and analysis of 222 variable stars and ten novas was unprecedented. The unusual nature of this achievement can be better appreciated when it is understood that most astronomers were proud to discover even one variable star and that they then typically left the further study of its properties to others. Fleming's influence on the many assistants who worked under her was even more significant. Her pioneering classification scheme led to major contributions by other women who followed her example at Harvard. Two of the most important of these were Annie Jump Cannon and Henrietta Leavitt, both of whom began as her assistants.

Annie Cannon joined the Harvard staff in 1896 and developed the definitive Harvard system of spectral classification based on the work of Fleming. She organized Fleming's alphabetical system according to increasing spectral complexity, and her system led to a later correlation with decreasing temperatures and an understanding of the evolution of stars. Henrietta Leavitt continued Fleming's study of variable stars and became a full staff member in 1902. Ten years later, she announced her discovery of the relation between the pulsation period and luminosity of a type of variable star called Cepheids that later led to measurements of the distances of galaxies and the discovery of the expansion of the universe.

Joseph L. Spradley

Further Reading

Belkora, Leila. *Minding the Heavens: The Story of Our Discovery of the Milky Way*. Bristol, England: Institute of Physics, 2003. Fleming is briefly discussed in a chapter on William Huggins and astronomical spectroscopy.

Cannon, Annie J. "Williamina Paton Fleming." *Astrophysical Journal* 34 (July, 1911): 314-317. Obituary of Fleming written by one of her colleagues at the Harvard College Observatory.

Dobson, Andrea K., and Katherine Bracher. "Urania's Heritage: A Historical Introduction to Women in

Astronomy." *Mercury: The Journal of the Astronomical Society of the Pacific* 21 (January/February, 1992): 4-15. This well-illustrated article includes a discussion of the work of Fleming and Cannon.

Jones, Bessie, and Lyle Boyd. *The Harvard College Observatory, 1839-1919.* Cambridge, Mass.: Harvard University Press, 1971. Several sections of this history discuss Fleming's work and contributions.

Kass-Simons, G., and Patricia Farnes, eds. *Women of Science: Righting the Record.* Bloomington: Indiana University Press, 1990. A chapter on "Women in Astronomy in America" has a good discussion of the Harvard College Observatory women.

Spradley, Joseph L. "The Industrious Mrs. Fleming." *Astronomy 18* (July, 1990): 48-51. A short account of Fleming's life and work with several photographs of her and others of the Harvard College Observatory women.

GALILEO

Italian astronomer and mathematician

Galileo helped establish the modern scientific method through his use of observation and experimentation. His work in mathematics, astronomy, and physics made him a leading figure of the early scientific revolution.

Born: February 15, 1564; Pisa, Republic of Florence (now in Italy)
Died: January 8, 1642; Arcetri, Republic of Florence (now in Italy)
Areas of achievement: Astronomy, mathematics, physics, science and technology

Early Life

Galileo (gahl-ih-LAY-oh) was the first of seven children born to Vincenzo Galilei and Giulia Ammanati. His father was a cloth merchant and a noted musician who wrote several treatises on musical theory. The Galileis were a noble Florentine family that over the years had lost much of its wealth. It was for financial reasons that Vincenzo left Florence and moved to Pisa to establish his textile trade.

At the age of ten, Galileo and his family returned to Florence. His early education was directed by his father with the help of a private tutor. He also spent some time at the monastery of Santa Maria di Vallombrosa. The content of Galileo's elementary education is unknown, but it was probably humanistic in character. His father urged him to pursue university studies that would lead to a lucrative profession.

Following his father's wish, Galileo enrolled as a student of medicine at the University of Pisa in 1581. He showed little interest in his medical studies; it was mathematics that captured his attention. A year after enrolling at the university, Galileo made his legendary discovery of the isochronal movement of pendulums by observing a chandelier in the Pisa cathedral. He confirmed his theory regarding the equal movement of pendulums by conducting a series of experiments. He continued an independent study of science and mathematics, finally convincing his father to allow him to abandon his medical studies. Galileo withdrew from the University of Pisa in 1586 without receiving a degree, and he returned to his family.

Upon his return to Florence, Galileo studied a wide range of literary and scientific texts. In addition, he delivered a series of popular lectures on the *Inferno* of Dante's *La divina commedia* (c. 1320; *The Divine Comedy*, 1802) at the Florentine Academy. In 1589, he used the influence of friends to obtain an appointment as a lecturer of mathematics at the University of Pisa. His return to Pisa marked a productive and enjoyable time for the young scholar. He conducted a series of experiments relating to falling bodies and wrote a short manuscript that challenged many traditional and generally accepted teachings about physics. In addition to his scholarly activities, Galileo was known for his quick wit, biting sense of humor, and excellent debating ability. Once again his friends intervened on his behalf to arrange an appointment, in 1592, to a more prestigious chair of mathematics at the University of Padua.

Life's Work

It was at Padua that Galileo began his life's work, which would bring him both fame and controversy. He quickly established himself as an excellent and popular teacher, both in terms of public lectures and private tutoring. He also wrote a series of short manuscripts on a variety of technical and practical issues. In 1597, he constructed a "military compass" to assist artillery bombardments and army formations.

Although Galileo's invention of the military compass brought him acclaim and a good source of additional income, it was his work in the study of motion and astronomy that firmly established his reputation as a leading

Galileo. (Library of Congress)

scientist. In 1604, a new star could be seen, and its sudden appearance prompted a fierce debate. According to the dominant theory of the time, Earth was the immovable center of the universe. Based on the work of Aristotle and Ptolemy, most scholars believed that the planets, Sun, and stars rotated around a stationary Earth. The universe was thought to reflect a perfect and unchangeable order that had been created by God. The new star raised a problem of how to account for its presence in an already complete and perfectly ordered universe.

The intensity of this debate reflected a larger controversy regarding the work of Nicolaus Copernicus. Copernicus claimed that Earth and the other planets orbited the Sun, and the stars were fixed or stationary. The appearance of a new star provided a tangible point of reference to settle a much larger scientific and theological debate on the structure and nature of the universe. It was a debate Galileo wanted to enter. As his correspondence with the astronomer Johannes Kepler indicated, Galileo found Copernicus's thesis convincing, but he lacked the necessary instrument to test the theory. This problem was remedied in 1609 when on a visit to Venice, Galileo learned about a new "eye-glass by means of which visible objects, though very distant from the eye, were distinctly seen as if nearby." Based on this limited information, Galileo returned to Padua to design and build his own telescope.

With this new instrument, Galileo turned his gaze toward the sky. He saw that the Moon was not a smooth sphere, as previously assumed, but had many craters and mountains. These geological characteristics implied that Earth was not a unique or central planet in the universe. While observing Jupiter, he discovered four moons, disproving the assumption that Earth was the only planet to be orbited by a natural satellite. His observations of Venus forced Galileo to conclude that its phases could not be accounted for within the traditional geocentric model of the universe but could only be explained in terms of the Copernican heliocentric system. His study of the Sun revealed its spots, implying that it was spherical and rotated on its axis as did Earth.

In 1610, Galileo published his findings in his *Sidereus nuncius* (*The Sidereal Messenger*, 1880). The book was quite popular and was translated and reprinted in a wide variety of languages. The book implied a strong support of the Copernican solar system. The vast number of stars and movement of planets Galileo observed could be explained only in the context of a heliocentric model wherein Earth and the other planets orbit the sun, and the stars are fixed points of light that appear to move only because of Earth's orbital path.

The Sidereal Messenger brought Galileo international fame and set the stage for future controversy. The University of Padua granted him a professorship for life. Instead of accepting the offer, however, he resigned so that he could return to Florence and become the grand duke's chief philosopher and mathematician. The move marked a fateful change for the scientist.

A year after his return to Florence, Galileo made a triumphant visit to Rome. He lectured widely, demonstrated his telescope, and debated a variety of scientific issues. Although he was well received by both the pope and papal court, there were signs of growing opposition to his work. Some theologians claimed that Copernicus, and therefore Galileo, was in conflict with the Bible and the doctrines of the Church regarding the central role and location of Earth in God's created order. Galileo was warned that he should teach and discuss the Copernican system only as a speculative theory and not as a truthful representation of the universe. In 1615, Copernicus's book, *De revolutionibus orbium coelestium* (1543; *On the Revolutions of the Heavenly Spheres*, 1939; better

Galileo on the Pendulum's Motion

Galileo used dialogue to present his ideas on motion in Dialogue Concerning the Two Chief World Systems, Ptolemaic and Copernican *(1632), a work banned by the Catholic Church. Two of the book's "interlocutors," Sagredo and Salviati, were close friends of Galileo, and here they set the stage for a discussion of the motion of a pendulum by wondering about the to-and-fro movement of a hanging church lamp.*

Sagredo. A thousand times I have given attention to oscillations, in particular those of lamps in some churches hanging from very long cords, inadvertently set in motion by someone, but the most that I ever got from such observations was the improbability of the opinion of many, . . . that motions of this kind are maintained and continued by the medium, that is, the air. It would seem to me that the air must have exquisite judgment and little else to do, consuming hours and hours in pushing back and forth a hanging weight with such regularity. . . .

Salviati. First of all, it is necessary to note that each pendulum has its own time of vibration, so limited and fixed in advance that it is impossible to move it in any other period than its own unique and natural one. Take in hand any string you like, to which a weight is attached, and try the best you can to increase or diminish the frequency of its vibrations; this will be a mere waste of effort. On the other hand, we [can] confer motion on any pendulum, though heavy and at rest, by merely blowing on it. This motion may be made quite large if we repeat our puffs; yet it will take place only in accord with the time appropriate to its oscillations. If at the first puff we shall have removed it half an inch from the vertical, by adding the second when, returned toward us, it would commence its second vibration, we confer a new motion on it. . . .

Source: Galileo, *Two New Sciences* (1638), translated and introduced by Stillman Drake (Madison: University of Wisconsin Press, 1974), pp. 98-99.

known as *De revolutionibus*, was placed on the Catholic Church's Index of Forbidden Books. In response to this censorship, Galileo refrained from any public comment on astronomy for a number of years, turning his attention to navigational problems. In 1618, however, three comets appeared, and a Jesuit astronomer maintained that their appearance disproved Copernicus. Galileo broke his silence with the publication of *Saggiatore* (1623, the assayer; partial English translation, 1990). The brief tract not only refuted the attack against Copernicus but also presented an elegant argument in behalf of free scientific inquiry.

As the controversy surrounding this episode subsided, Galileo began writing his most important and controversial book, *Dialogo sopra i due massimi sistemi del mondo, tolemaico e copernicano* (1632; *Dialogue Concerning the Two Chief World Systems, Ptolemaic and Copernican*, 1661). Its purpose was to present an "inconclusive" comparison between the Ptolemaic and Copernican models of the universe. Although Galileo carefully presented the various claims in terms of competing theories, it was apparent that he believed that the Ptolemaic geocentric theory was false and that the Copernican heliocentric theory was true. Although the book had received the Catholic Church's imprimatur, it was placed on the Index of Forbidden Books shortly after its publication. A few months later, the Office of the Inquisition summoned Galileo to Rome to stand trial for heresy. What was at stake was whether he had defied a papal ban "to hold, defend, and teach the Copernican doctrine." More important, the Church's authority and ability to enforce compliance with its teachings were also at issue.

After a five-month trial, Galileo was convicted because he "held and believed false doctrine, contrary to the Holy and Divine Scriptures." The punishment would be a public prohibition of *Dialogue Concerning the Two Chief World Systems* and a prison sentence. Galileo, however, was given an opportunity to recant, which he accepted, swearing, "I will never again say or assert . . . anything that might furnish occasion for a similar suspicion." Theological authority, for the time being, had silenced the claims of scientific observation.

Significance

For the rest of his life Galileo remained under house arrest, first in the village of Siena and later in Arcetri. He was not allowed to take any extensive trips or to entertain many guests. Following the death of his favorite daughter in 1634, he lived a lonely life and became blind in 1637. Despite the attempt to isolate him from the world, his fame grew—such noted figures as Thomas Hobbes and John Milton went out of their way to visit him shortly before his death.

His legacy was to establish science, based on observation and experimentation, as an important intellectual and social force in the world. Unlike Copernicus or Kepler, he was not a systematic or speculative thinker, preferring to base his work on a careful inquiry into the causes of natural phenomena. As indicated by his various inventions, he also was interested in applying his knowledge to practical problems.

Galileo marked an important break between theology and science that was not easily or quickly bridged—Copernicus and he were not removed from the Church's Index of Forbidden Books until 1835. Yet despite his conviction for heresy, history has judged his right to seek after truth quite differently. According to legend, as Galileo signed his recantation following his trial, he mumbled, *"Eppur si muove"* (and yet it [Earth] moves). Although Galileo never spoke these words, the legend's existence and endurance is a fitting indication of the eventual support he received not only for his work but also for the right of the scientist to engage in free and open inquiry.

Brent Waters

Further Reading

Boas, Marie. *The Scientific Renaissance, 1450-1630*. New York: Harper & Row, 1962. A detailed historical account of the major scientific discoveries and conflicts during the Renaissance. Provides good background material regarding the work and accomplishments of Galileo.

Fermi, Laura, and Gilberto Bernardini. *Galileo and the Scientific Revolution*. New York: Basic Books, 1961. A concise and highly sympathetic biography. Includes a limited number of illustrations and a translation of Galileo's first tract, "The Little Balance."

Freedburg, David. *The Eye of the Lynx: Galileo, His Friends, and the Beginnings of Modern Natural History*. Chicago: University of Chicago Press, 2002. Galileo was a member of the Academy of Lynxes, a group of scientists who edited and published his work. This book describes how the group used microscopes and other tools to study and illustrate plants and animals, aiming to represent all of nature in pictures.

Geymonat, Ludovico. *Galileo Galilei*. New York: McGraw-Hill, 1965. A highly detailed biography focusing on the development of Galileo's thinking. Particular attention is placed on reconstructing his emerging philosophy of science.

Koyré, Alexandre. *Galileo Studies*. Translated by J. Mepham. Atlantic Highlands, N.J.: Humanities Press, 1978. A highly technical and critical examination of the scientific and mathematical principles used by Galileo in his various observations and experimentations.

Kuhn, Thomas S. *The Copernican Revolution: Planetary Astronomy in the Development of Western Thought*. Cambridge, Mass.: Harvard University Press, 1957. A detailed historical review of the debate inspired by Copernicus. The intellectual and social implications of the change from a geocentric to a heliocentric worldview are also examined.

Redondi, Pietro. *Galileo: Heretic*. Translated by Raymond Rosenthal. Princeton, N.J.: Princeton University Press, 1987. A comprehensive and critical examination of Galileo's trial before the Inquisition. Particular attention is directed toward examining the motivations and issues at stake for the Catholic Church.

Reston, James, Jr. *Galileo: A Life*. New York: HarperCollins, 1994. A character study and political biography, written in the present tense. The major portion of the book is devoted to Galileo's trial.

Ronan, Colin A. *Galileo*. New York: G. P. Putnam's Sons, 1974. A standard biography that not only provides numerous details about Galileo's life but also places them within the larger context of the intellectual changes taking place. Contains numerous illustrations and photographs.

Rowland, Wade. *Galileo's Mistake: A New Look at the Epic Confrontation Between Galileo and the Church*. New York: Arcade, 2003. Rowland argues that Galileo and church officials disagreed about something more significant than whether Earth revolved around the sun; they were disputing the nature of truth and how people acquire the truth.

GEORGE GAMOW

Russian-born American physicist

A member of the select group of physicists who developed quantum theory, applied it to the nucleus, and moved on to radiation, Gamow also turned to astrophysics and promoted the big bang theory of the origin of the universe. He predicted the existence of background radiation left over from that singular event. He advanced an explanation for why the stars shine, how the elements were created, and a method of analyzing deoxyribonucleic acid, or DNA.

Born: March 4, 1904; Odessa, Russia (now in Ukraine)
Died: August 20, 1968; Boulder, Colorado
Also known as: Georgy Antonovich Gamov (full name)
Areas of achievement: Physics, astronomy, genetics, literature

Early Life

George Gamow (GAM-awf) was the son of Anton Gamow, a teacher of Russian language and literature, and Alexandra Lebedinzev, whose father was Metropolitan Bishop of Novorossia. A big, flaxen-haired boy, young Gamow had an impish sense of humor, a gift for languages, and a lively intellect. The Odessa he grew up in was a part of the Russian Empire, and the great changes that the Soviet Union brought would eventually drive Gamow out of Russia.

In his youth, Gamow was fascinated by science, particularly astronomy. He attended Novorossia University in Odessa from 1922 to 1923, studying in the physico-mathematical faculty. Gamow left home in 1923 for the University of Leningrad, a move financed by the sale of the family silver. Leningrad was a major center of modern physics, and Gamow studied with Alexander Friedman, who postulated that the universe was expanding. Gamow's forward progress was delayed a bit by two problems. One was his inability to carry out experiments, and the other was an exceptionally dull dissertation topic he simply could not force himself to address. A professor suggested Gamow might profit from a few months at a foreign university, and so, in 1926, Gamow found himself at Göttingen. It was a seminal stay. Although he had not yet received his doctorate, Gamow's obvious brilliance allowed him to begin a career of great distinction.

Life's Work

Göttingen was the beating heart of the quantum revolution that was overtaking physics, and Gamow discovered a paper in which the British physicist Ernest Rutherford presented a puzzling behavior, and question. Why, he asked, did high-energy alpha particles (later found to be helium nuclei), when fired at a uranium nucleus from outside, bounce off quite rapidly, when alpha particles that were inside the uranium nucleus were able, although slowly, to stream out of the nucleus with no external perturbation?

Gamow decided to tackle this problem using quantum mechanics; in so doing, he treated the escaping alpha particle as a wave-particle, not simply as a particle. The dual nature of the wave-particle allows the possibility of movement through an energy barrier. Gamow's solution came to be called quantum tunneling, which is an important consequence of powerful wave mechanics. He had also thereby laid the groundwork for a successful dissertation.

The next step was a Carlsberg Fellowship in Copenhagen with Niels Bohr. The few months he had off turned into a year and a half. A brief and very pleasant visit home to Russia, including a visit with his aging father, was followed by a year with Rutherford at the Cavendish Laboratory at Cambridge University. Here he once again focused on the atomic nucleus, and he spent another six months at Copenhagen with Bohr. However, while Gamow was learning the new physics, changes were going on inside the Soviet Union, of which he was apparently oblivious. When he headed home, he expected that nothing much would have changed. However, he was to be unpleasantly surprised.

Gamow's friends could not believe he had returned to the Soviet Union. In the spirit of dialectical materialism, the Communists were in full attack mode against capitalistic science and scientists. Gamow's working knowledge was judged to be an untrue science. Many of his friends and some of the people he admired were being taken away by the government. Although he was scheduled to attend an international conference, Gamow was not allowed to leave the Soviet Union, and his passport was confiscated.

Gamow's spirits were buoyed by meeting a young woman who also was working in physics: Lyubov Vokhminzeva (nicknamed Rho); the two were soon married. It quickly became clear that Gamow would be interned in a concentration camp for his scientific work and his ideas of the natural world. The Gamows made two rather reckless attempts at escape, but both attempts failed.

In the fall of 1933, Gamow received a letter from Soviet government officials. To his astonishment, he was ordered to go to Brussels to represent the Soviet Union at the Solvay Congress on nuclear physics in October. Knowing that he would probably never go back to the Soviet Union once outside the country, he refused to leave without his wife. An old friend provided two passports so that the couple could travel together to the congress in Brussels; they had no money and no known means of support. The scientific community acted swiftly. In short order, Gamow would receive brief research jobs, one in Paris at the Pierre Curie Institute, another in Copenhagen with Bohr, and another in Cambridge with Rutherford. Next was a summer teaching position in Ann Arbor, Michigan, and then a permanent professorship at George Washington University in Washington, D.C. The couple's only child, Rustem Igor, was born at George Washington in 1935.

Once secure, Gamow showed that he was able to work at a swift pace. A paper he cowrote with doctoral student Ralph Alpher on the origin of chemical elements supported the then-contentious and novel theory of the big bang. Gamow and Alpher, along with another doctoral student, Robert Herman, would estimate the strength of cosmic microwave radiation remaining from the big bang after the billions of years of cooling, an estimate forgotten in the 1950's. In 1964, however, the strength of the remaining cosmic microwave radiation was shown to be close to correct. Furthermore, Gamow worked out the life cycle of stars and a set of equations for figuring the mass and the radius of a primordial galaxy. The thermonuclear fusion that makes the bright light of stars gave Gamow an understanding of these reactions that led to the development of the hydrogen bomb in the United States after World War II.

Given the time period, one might expect Gamow would have worked on the Manhattan Project during World War II, but because he was a colonel in the Red Army, he could not get a security clearance from the U.S. government. Somewhat frustrated, he consulted for the U.S. Navy. One of his tasks was to bring to Albert Einstein a weekly load of secret papers. Every week, Gamow and Einstein went through the Navy's proposed projects together. Gamow received a security clearance after the war, and he participated in the further development of atomic weapons at Los Alamos Laboratory in New Mexico.

James D. Watson and Francis Crick's discovery (with the critical research of Rosalind Franklin) of deoxyribonucleic acid (DNA) in 1953 enlarged the scope of Gamow's work. In 1954, he was a visiting professor at the University of California, Berkeley, when a friend told him that Watson and Crick had written an article that laid out the structure of DNA. Gamow, who was attracted as always by a puzzle, saw that the sequence of nucleotides formed a sort of code. One year later, he proposed a theory related to the organization of genetic information based on this code. His theory allowed biochemists to begin work on discovering how the four nucleotides, arranged three at a time, could specify twenty amino acids.

The path less taken in Gamow's professional life by his interest in DNA was matched by a major change in his personal life and in his university affiliation. In 1953, Gamow met Barbara Perkins, who handled the marketing of his books. By 1956, he had divorced Vokhminzeva, and in 1958, he married Perkins and moved to the University of Colorado. He continued to write, developing his charming popular science series featuring the odd character Mr. Tompkins, whose fantastic adventures illustrated difficult scientific theories. Gamow continued his scientific work, accepting a fellowship at Churchill College, Cambridge, in England in 1965. An unspecified illness led to his death on August 20, 1968, in Boulder, Colorado.

Significance

Gamow's work in quantum theory, the nuclei of atoms, the beginnings of the universe, and the life of stars was critical to astronomy and physics. His intuition was powerful, as his analysis of DNA sequences shows. His efforts, however, were slightly out of phase with work being done by others. For example, he had failed to publish first his original idea, that of the big bang, and so is considered a supporter, and not the creator, of the big bang theory. It seems more and more likely that Gamow, brilliant as he was, will be remembered as an influential but minor figure on the world stage of physics in the twentieth century.

In hindsight, Gamow's significance might be in his popularization of science. His Mr. Tompkins series is fun, easy to read, and informative. His more scholarly works are accessible to educated readers. His quirky autobiography *My World Line* (1970) is a wonderful introduction to his life.

Jane Carter Webb

Further Reading

Gamow, George. *Mr. Tompkins Gets Serious: The Essential George Gamow*. Edited by Robert Oerter. New York: Pi Press, 2006. A Masterpiece Science edition, which includes a foreword by Gamow's son, Igor Gamow. A collection of essential works from Gamow's Mr. Tompkins science series.

_______. *Mr. Tompkins in Wonderland* and *Mr. Tompkins Explores the Atom*. New York: Cambridge University Press, 1993. The first title stars an absentminded bank clerk whose tendency to fall asleep brings unusually strange dreams. In the second title, peculiar adventures abound as a dreamy Tompkins turns into an electron, only to be annihilated by a chance encounter with another electron. Tompkins discovers the second law of thermodynamics and the rules of gaming, and he falls in love with Maud.

_______. *Mr. Tompkins Learns the Facts of Life*. New York: Cambridge University Press, 1953. In this quirky tale, Gamow tells the story of how cells combine into molecules and how life blossoms.

_______. *Thirty Years That Shook Physics: The Study of Quantum Theory*. 1966. Reprint. New York: Dover, 1985. An introduction to quantum mechanics written by a great expositor of the theory. This book contains the text of a quantum mechanical Faust performed at Niels Bohr's institute. A good starting point for those unfamiliar with modern physics.

Gamow, George, with Martynas Ycas. *Mr. Tompkins Inside Himself: Adventures in the New Biology*. New York: Viking Press, 1967. This further exploration of biology is carried on by the son of the characters Tompkins and Maud, a tiny Tompkins who struggles through his own internal anatomy. Includes a foreword by Roger Penrose.

Greenstein, George. *Portraits of Discovery: Profiles in Scientific Genius*. New York: John Wiley, 1998. Includes a biographical essay on Gamow "the Magician."

Stuewer, Roger. "Gamow." *Dictionary of Scientific Biography*. Vols. 5-6. New York: Charles Scribner and Sons, 1981. An excellent, lengthy biographical article on Gamow.

Yockey, Hubert P. *Information Theory, Evolution, and the Origin of Life*. New York: Cambridge University Press, 2005. Includes the chapter "James Watson, Francis Crick, George Gamow, and the Genetic Code," examining the role of all three scientists in the discovery of DNA and the implications of this discovery for information theory in biology. For advanced students in the sciences.

GEORGE ELLERY HALE

American astronomer

Hale was instrumental in constructing the world's largest telescopes of the early twentieth century. He also constructed two towers for solar telescopes, using them to discover that sunspots are associated with intense magnetic fields. He was the first to show that magnetic activity cycles were the key to understanding solar activity, a discovery that marked the beginning of solar astrophysics.

Born: June 29, 1868; Chicago, Illinois
Died: February 21, 1938; Pasadena, California
Areas of achievement: Astronomy, invention and technology

Early Life

George Ellery Hale was the third child born to William Ellery and Mary Scranton Hale, but he was the first child to survive. Hale started school at Oakland Public School, but his parents switched him to the private Allen Academy after a bout with typhoid. At age fourteen, Hale began studies at the Chicago Manual Training School. Many summers were spent at his maternal grandmother's home in Madison, Connecticut. There, Hale met and began dating Evelina Conklin. In 1886, Hale entered the Massachusetts Institute of Technology (MIT) in Cambridge to study physics. At about the same time, he announced his engagement to Conklin. They were married in June of 1890, shortly after Hale's graduation.

Hale became extremely interested in science early on, and his father supported and encouraged his interests. Hale's interest in astronomy was fostered by a neighbor, a famous double-star astronomer named Sherburne W. Burnham. Burnham showed Hale a spectrum of the sun, beginning an interest in solar studies that would remain with him for the rest of his life. In 1884, at the age of sixteen, Hale used a prism spectroscope and telescope purchased by his father to photograph the sun's spectrum.

Life's Work

Even as an undergraduate at MIT, Hale was doing research on the sun. Still an undergraduate, he invented the spectroheliograph, an instrument that could image the sun using only one wavelength of light, the hydrogen alpha line. The instrument enabled astronomers to study the sun's chromosphere and to photograph solar prominences.

In 1891, Hale established a small private observatory near Chicago named the Kenwood Observatory, where he continued his studies. Later in life, Hale would think back fondly of his time at Kenwood. In 1892, he was hired by the University of Chicago as an associate professor of astrophysics, the first person to ever hold that title. Shortly after being hired at Chicago, Hale learned of two very large lens blanks produced by the renowned telescope maker Alvan Clark. He began at once working to secure funding for the University of Chicago to build an observatory with the largest refracting telescope ever constructed. Together with university president William Harper, Hale managed to persuade the wealthy streetcar industrialist Charles Yerkes to finance construction of an observatory for the university near Williams Bay, Wisconsin.

Hale commissioned the famous optics company Alvan Clark & Sons to grind the lenses for a giant refracting telescope 40 inches in diameter as the centerpiece for the observatory. After several cost overruns, which required that Hale persuade Yerkes to donate even more money, the Yerkes Observatory opened in 1897. Hale hired Edward E. Barnard, Frank Wadsworth, and his old mentor, Burnham, to staff the new observatory. Assisting these astronomers were Ferdinand Ellerman and George W. Ritchey, with Walter Adams joining them later. The primary astronomers were already well known, but they all came to be powerhouses in the field of astronomy over their careers.

Hale did more at Chicago than just build the observatory. Together with J. Edward Keeler, Hale founded *The Astrophysical Journal* in 1894, a journal that would become a premier astronomical publication. In 1899, Hale founded the first professional organization for American astronomers. This organization, the American Astronomical and Astrophysical Society, later changed its name to the American Astronomical Society.

In no time Hale wanted to build an even larger telescope. With Ritchey he realized that the Yerkes refractor was just about the limit in terms of how large lenses could be made for telescopes. Larger telescopes would have to be reflectors. Indeed, the Yerkes refractor was the last large refracting telescope. Hale also realized that the ideal site for a new far larger telescope would be on top of a mountain, where steady clear air and minimal light pollution would facilitate observations. Hale traveled to Southern California to find a site for a new, much larger telescope.

Wilson's Peak (now known as Mount Wilson), near Pasadena, California, was chosen as the site. Hale and Ellerman set up a small 5-inch solar telescope on the mountain to establish a foothold for an observatory there. In 1904, Hale received a grant from the Carnegie Institution of Washington to establish the Wilson Observatory. The grant allowed Hale to begin work on a telescope with a 60-inch mirror. By 1905, he had located a 24-inch solar telescope, the Snow telescope, on Mount Wilson and had photographed for the first time the spectrum of a sunspot. The 60-inch telescope, the largest in the world at that time, began operation in 1908.

By 1912, Hale had also constructed two towers for solar telescopes, and had used them to discover that sunspots are associated with extremely intense magnetic fields. He was the first to realize that magnetic activity cycles were the key to understanding solar activity. This marked the beginning of serious solar astrophysics.

Even as the 60-inch telescope was under construction, Hale was yearning for a larger telescope. He had managed to secure funding from John D. Hooker, a wealthy California businessman, for the purchase of a mirror blank 100 inches in diameter. By 1910, Hale had convinced Andrew Carnegie to provide another grant to figure the 100-inch mirror and to construct the telescope. The 100-inch telescope, later named the Hooker telescope, was finally completed in 1917, and it remained the world's largest for the next three decades. Researchers who used the telescope made some of the most important astronomical discoveries of the early twentieth century.

The design and construction of the Hooker telescope, though, took its toll on Hale. He had always been seen as somewhat somber, but he drove himself mercilessly. By 1909, he was showing signs of severe depression, possibly even bipolar disorder (diagnoses that did not exist at the time). In 1910 he had what was termed a "nervous breakdown." On his doctor's orders, he had to take a rest. Hale was unable to work for nearly a year, though he did keep abreast of the figuring of the 100-inch mirror. By 1912, Hale was again able to serve as director of the Mount Wilson Observatory.

By 1914, World War I had begun. Hale, realizing that the United States would eventually be part of the war and that technological superiority would play a role in winning the war, helped to establish the National Research Council, a group organized by the National Academy of Sciences in 1916. The council, with Hale as its first president, organized the war efforts of the nation's scientists.

After the war, Hale had several episodes of severe depression. He eventually had to limit himself to a few hours of work per day, on those days he could work, eventually retiring as director of the Mount Wilson Observatory in 1923. He then retreated to his own private solar observatory, where he worked in isolation. However, he did not completely remove himself from the astronomical community. By 1928, he had already launched plans for what would be the largest telescope in the world: a 200-inch-diameter reflector. He secured initial funding from the Rockefeller Foundation to site the instrument atop Mount Palomar in California. By 1932, though, his depression returned in full force. He spent the rest of his life in and out of sanitaria (mental hospitals) and his final year and a half in a nursing home. World War II delayed construction of the 200-inch telescope at Palomar, but when it was finally completed in 1948, it was named the Hale telescope.

Significance

Hale made major strides in solar astronomy, developing instruments and techniques that are still in use. His observations of the sun set the stage for all solar astronomers seeking to understand its composition.

Hale's vision and drive to build new and bigger telescopes was foundational in making the United States a major contender in the science of astronomy. The Yerkes refractor remains the largest refracting telescope ever constructed.

Raymond D. Benge, Jr.

Further Reading

Glass, I. S. *Revolutionaries of the Cosmos: The Astro-Physicists*. New York: Oxford University Press, 2006. An excellent book of biographies of some of the most important astronomers. A chapter is devoted to Hale.

Osterbrock, Donald E. *Pauper and Prince: Ritchey, Hale, and Big American Telescopes*. Tuscon: University of Arizona Press, 1993. A chronicle of Hale's observatory-building activities.

_______. *Yerkes Observatory, 1892-1950: The Birth, Near Death, and Resurrection of a Scientific Research Institution*. Chicago: University of Chicago Press, 1999. The story of Hale's first major observatory.

Sandage, Allan. "The Mount Wilson Observatory." In *Centennial History of the Carnegie Institution of Washington*. New York: Cambridge University Press, 2004. A look at the Wilson Observatory, funded in part by the Carnegie Institution, and its first one hundred years of operation.

Tenn, Joseph S. "George Ellery Hale: The Thirteenth Bruce Medalist" *Mercury* 21, no. 3 (May/June, 1992): 94-97. An excellent, brief biography of Hale and his life's work.

EDMOND HALLEY

English astronomer

Among his many scientific achievements, Halley's best-known accomplishment was to solve the riddle of the orbits of comets. In particular, he predicted that the one seen in 1682 would return in 1759. This comet was later named for him.

Born: November 8, 1656; Haggerston, near London, Shoreditch, England
Died: June 14, 1742; Greenwich, near London, England
Areas of achievement: Astronomy, science and technology

Early Life

The details of the life of Edmond Halley (HAL-ee) are only sketchily known, since most of his private papers and correspondence have been lost. Some facts have been collected from the papers of colleagues, relations, and friends, but many gaps remain. It is known that he was born on November 8, 1656, in Haggerston, on the outskirts of London. His father, a soap boiler, belonged to the monied merchant class and provided for Halley throughout his education. Halley attended St. Paul's School, where he became the school captain, and furthered his studies at Queen's College, Oxford, in 1673. It was there that he first formally studied astronomy, beginning a regular correspondence with John Flamsteed, who was then astronomer royal. The older man took the young Halley under his wing, and with Flamsteed's guidance, Halley made one or two minor contributions to the mathematics of astronomy.

In 1676, Halley ventured to Saint Helena in the Southern Hemisphere on an expedition to study the stars there. He was supported during this time by his father, who was still comparatively wealthy, despite having suffered some property losses during the 1666 Great Fire of London. Halley had an allowance of three hundred pounds a year. Although conditions at Saint Helena were overcast, he achieved his aim and on his return published the *Catalogus Stellarum Australium* (1678; Australian stellar catalog) and a planisphere. His work found favor with King Charles II, since Halley regrouped a particular constellation of stars and renamed them Charles's Oak in his honor.

Halley was awarded an M.A. at Oxford for this work, and his endeavors brought him to the attention of the Royal Society (the somewhat august body to which all the leading scientists of the day belonged). The society made him a fellow. He was asked by the Royal Society to visit the Dutch astronomer Johannes Hevelius at Danzig to investigate his methods of star observation. With telescopic sights now in vogue, the society believed that Hevelius's methods were old-fashioned and therefore inaccurate. Halley reported in Hevelius's favor (much to Flamsteed's annoyance and causing a breach of friendship between the two that lasted some forty years).

While Halley's scientific reputation soared with these projects, however, his personal reputation was surrounded by rumor and innuendo. On the Saint Helena expedition, his name was coupled with that of a married woman who found herself pregnant on Halley's departure, and during the Danzig trip, gossip abounded about Halley's relationship with Elisatetha Hevelius, Johannes's wife. In the coffeehouses, Halley appears to have been labeled a womanizer, and his reputation offended some of his more straitlaced colleagues.

In 1680, Halley traveled to Paris to work with Gian Domenico Cassini, the director of the Paris Observatory. The two made detailed observations of a new comet that had appeared. This was not the comet that would later bear his name, but this kind of painstaking observational work was to be Halley's hallmark throughout his scientific career. He returned to London, perhaps because his father could no longer support him, and married Mary Tooke in 1682. They settled in Islington, London, and remained married until her death in 1736. Little is known about her or about their life together.

Life's Work

The next period of Halley's life was bound up with his own meticulous observations, the continual wranglings of the Royal Society, and his friendship with Sir Isaac Newton, undoubtedly the greatest scientist of that generation. The Royal Society at that time contained some of the most prominent men of the seventeenth century—Sir Christopher Wren, Samuel Pepys, Robert Hooke, Flamsteed, and Newton—and the papers and theories that were presented at its meetings were in the vanguard of scientific research.

The late seventeenth century was an exciting period in the history of astronomy. Following the pioneering observational work of Galileo and Johannes Kepler, planets were known to travel in regular, elliptical orbits around the sun. The problem facing the scientists of the Royal Society was to explain the mechanics of this occurrence, demonstrating from first principles, mathematically, why it should be so. Many of the eminent men of the day were tackling the challenge, and the race was on to reach the solution first. Halley played no small part in this story.

In 1684, Halley became the clerk of the Royal Society, a post that was later accompanied by a stipend of fifty pounds a year, although there is some evidence to suggest that he rarely actually got paid. This post involved much organizational work for the society, but it also placed Halley in the thick of its debates. Around this time, wanting to confirm some of his observations on comets, Halley went to visit Newton at Cambridge. During their discussions, Newton claimed to have solved the elliptical problem, asserting that the elliptical shape of planetary orbits was a result of the inverse square law of attraction, which states that bodies are attracted to one another in

Edmond Halley (Library of Congress)

proportion of the inverse square of their distance from one another. Newton said, however, that he had mislaid the proof, so Halley was not totally convinced.

On Halley's second visit, Newton successfully produced a nine-page paper explaining the elliptical orbits of the planets, and Halley, recognizing the overwhelming importance of Newton's work, persuaded him to publish it. Through Halley's offices, Newton published first a short tract, *De motu corporum in gyrum* (1684; on the motion of bodies in an orbit), which demonstrated Kepler's laws of planetary motion, and then his famous book *Philosophiae Naturalis Principia Mathematica* (1687; *Mathematical Principles of Natural Philosophy*, 1729), in which he elaborated his three laws of mechanics and his theory of gravity.

This work was, perhaps, the most important scientific work to be published in the century and, throughout, Halley was Newton's friend, confidant, editor, and publisher, encouraging him to continue, adding the occasional helpful comment to the manuscript, and even finally paying for its publication. He also had the difficult and diplomatic task of mediating between Newton and Hooke, who were rivals, the latter claiming that Newton had not credited his contribution to the discovery of the inverse square law. Halley's involvement in this dispute made Hooke his enemy too, and Hooke tried to remove him from his post of clerk to the Royal Society. Halley's efficiency and popularity won the day, however, and Hooke's motion was defeated.

During this period of his life, Halley, who had numerous diverse interests, wrote a paper trying to explain the causes of the biblical Flood. This paper was to have disastrous consequences for his career. He argued that the Flood had occurred for natural, not miraculous reasons, possibly from a close encounter with a comet. (It is interesting to note that speculation about the effects of comets on the Earth has flourished ever since.) The result of Halley's reasoning was to place the date of the Flood somewhat earlier than the accepted orthodoxy of 4004 B.C.E.

Halley's theory of the Flood offended his more religious colleagues, and rumors abounded that Halley might be a skeptic or, worse still, an agnostic. In 1691, a vacancy opened for the Savilian Chair of Astronomy at Oxford. Were it not for this paper and the question concerning Halley's religious beliefs, he would certainly have been appointed. Flamsteed, however, argued vehemently against him, and not even Newton came to his protégé's defense. Later, in 1694, when he finally presented the offending paper to the Royal Society, he added the explanation that he was not actually referring to the Flood itself but speculating on what might have happened given the proper circumstances—thus mitigating some of the damage to his reputation.

Although Halley is best known for his work in astronomy, he made contributions in a number of other areas during his life, perhaps motivated by his persistent lack of money. He was interested in nautical problems, designing various practical instruments: a prototype diving bell, a device for measuring the path of a ship, and another device for measuring the rate of evaporation of seawater. This latter device had consequences for another interest of his, chronology. Indeed, it led to some of his conclusions about the date of the Flood.

Between 1696 and 1698, Halley's career took another turn. He became the deputy comptroller of the Royal Mint at Chester, a post offered him by Newton, who was then the warden of the Mint. (At the time, there was much trouble with "clipped" gold and silver coins, and scientists were being used to supervise the minting.) Administration did not prove to be one of Halley's many talents, however, and Newton found himself having to defend his friend against the Lord's Commissioners. In 1698, Halley set out on another expedition, this time to the South Seas to study the magnetic variations of the earth's compass. The journey was abandoned (with the ship's first lieutenant facing a court-martial on their return), but Halley tried again a year later with more success. He also went on a secret mission in 1701, about which little is known, traveling to France for the Admiralty on the pretext of yet another scientific expedition.

While Halley's accomplishments were many, the achievements for which he is best known are in the field of astronomy, particularly regarding the orbits of comets. This was a field on which he had already begun detailed work when he first met Newton. By 1695, he had undergone a thorough search of historical records on the subject, examining the figures with scrupulous attention. Recognizing that comets, too, were affected by the gravitational influence of the planets (following Newton's work), Halley calculated their possible orbits from the available data. That led him to disagree with Newton, who, in book 3 of *Mathematical Principles of Natural Philosophy*, had argued that comets' orbits should be parabolic. Halley concluded that they must, in fact, be elliptical like the planets themselves.

In 1703, Halley became a member of the Council of the Royal Society in recognition of his work, and in the same year, he was appointed to the Savilian Chair of Geometry at Oxford, previous doubts about his religion being now forgotten. Two years later, he published *Astronomie cometicae synopsis* (1705; synopsis of cometary astronomy), in which he outlined his theory of elliptical orbits for comets and asserted that the particular comet seen in 1682 would, as part of its periodic cycle, return in 1758. Although he did not live long enough to test his hypothesis, when the comet did come into view in 1759 (Halley made a minor miscalculation), it was seen not only as a verification of Halley's prediction but also as a major confirmation of the truth of Newton's mechanics. The comet was named for Halley. In 1719, on Flamsteed's death, Halley succeeded to the post of astronomer royal, a position he held until his death in 1742.

Significance

Halley lived at a time when the old theories of the structure of the universe, which were still based on the teachings of Aristotle, were being overturned. Experimentally, Galileo and Kepler had led the way, but it took the work of Newton to provide a theoretical underpinning to it all. Halley, in publishing Newton's work, had been largely responsible for bringing it to the attention of scientists and the public. Newton's theories turned the scientific world upside down, causing debate, discussion, and disagreement for nearly a century after their publication. One of the major triumphs of the new theories, helping eventually to bring about their universal acceptance, was the reappearance of Halley's comet, which he had predicted using Newton's work and his own observations. Halley's meticulous recording of observational detail and his ready grasp of contemporary theories had enabled him to become the champion of Newtonian science. In this way, he was central to one of the most exciting periods of scientific history.

Sally Hibbin

Further Reading

Armitage, Angus. *Edmond Halley*. Camden, N.J.: Thomas Nelson and Sons, 1966. A detailed historical evaluation of Halley's scientific research, mainly based on the papers he contributed to *Philosophical Transactions*, with relevant biographical information. Illustrated with plates and figures.

Calder, Nigel. *Comets: Speculation and Discovery*. New York: Dover, 1994. A popular account of the science, facts, and legends related to Halley's comet, and the people involved in its discovery and observation. Originally published in 1981 as *The Comet Is Coming! The Feverish Legacy of Mr. Halley*.

Cook, Alan. *Edmond Halley: Charting the Heavens and the Seas*. New York: Oxford University Press, 1998. A full-length biography drawing upon recently acquired information to discuss Halley's life and his contributions to science.

Halley, Edmond. *Correspondence and Papers of Edmond Halley*. Edited by Eugene Fairfield MacPike. Oxford, England: Clarendon Press, 1932. Reprint. New York: Arno Press, 1975. An edited collection of Halley's surviving papers and letters, with a memoir of his life by one of his contemporaries, and the "Éloge" by d'Ortous de Mairan. The best, although necessarily incomplete, primary source for students of Halley.

Lancaster Brown, Peter. *Halley and His Comet*. Poole, Dorset, England: Blandford Press, 1985. Occasioned by the anticipated appearance of Halley's comet in 1986, a modern reappraisal of Halley's life and work, its relevance to the origins of life on Earth, and its significance to the space age. Highly informative for specialists and lay readers.

Newton, Sir Isaac. *A Dissertation on Comets*. London: c. 1750. Describes the mechanical principles of the motions of planets, using Newton's new theory of gravity, from which Halley developed his own theory that comets travel in ellipses and will therefore return.

________. *The Mathematical Principles of Natural Philosophy*. Translated by Andrew Motte. London: B. Motte, 1729. Originally published by Halley, who, recognizing the importance of Newton's theories of gravity and motion, persuaded him to put it into book form. Motte's translation remains the standard text.

Thrower, Norman J. W., ed. *Standing on the Shoulders of Giants: A Longer View of Newton and Halley*. Berkeley: University of California Press, 1990. A collection of essays by astronomers, historians of science, and other writers that re-examine the professional relationship of Newton and Halley, describing their influence upon each other and on subsequent generations of scientists.

WILLIAM ROWAN HAMILTON

Irish mathematician

While questioning a commonly accepted three-dimensional concept of space on a plane, Hamilton discovered quaternions and, in doing so, drastically altered the study of algebra, forcing the abandonment of the commutative law of multiplication that was dominant in his day and leading the way to new methods of vector analysis.

Born: August 3/4, 1805; Dublin, Ireland
Died: September 2, 1865; near Dublin, Ireland
Area of achievement: Mathematics

Early Life

William Rowan Hamilton was born exactly at midnight, a moment poised equally between August 3 and 4, 1805. His father, Archibald Hamilton, was away in the north at the time of his son's birth, carrying out his duties as agent to Archibald Rowan, a post he had held since 1800. Archibald Rowan, who was William Rowan Hamilton's godfather, had been in exile for eleven years. His agent, William's father, worked tirelessly to make possible Rowan's return to his estate at Killyleagh, an effort that resulted in Rowan's repatriation in 1806.

To help Rowan meet his expenses, Archibald Hamilton borrowed heavily at high interest rates. When these loans were called, Rowan failed to back Hamilton, who, in a year or two, had no alternative but to declare bankruptcy. By 1808, the family was sufficiently impoverished not to be able to provide for William, then three years old, and his sisters, Grace and Eliza, who had to be sent away to be cared for by relatives. The two girls presumably were sent to live with their father's sister, Sydney, and young William became the ward of his uncle, James Hamilton, a Church of England clergyman who ran the diocesan school at Trim, some forty miles to the northwest of Dublin in County Meath. William was to remain there until 1823, when he returned to Dublin as a student at Trinity College.

In retrospect, it appears to have been a stroke of good fortune that young William was forced by circumstance to live with his uncle, a man of considerable intellect. Before the boy was four years old, he was able to read English and showed a remarkable understanding of arithmetic. By the time he was five, William was able to translate from Latin, Greek, and Hebrew. He knew Greek and Latin authors well enough to recite from their works, and he was also able to recite passages from works by John Milton and John Dryden. He is said to have mastered fourteen languages by the time he was thirteen. Before he turned twelve, he had compiled a Syriac grammar, and two years later he was sufficiently fluent in Persian to compose a speech of welcome that was delivered to the Persian ambassador when he was a guest in Dublin.

Always advanced in mathematics, Hamilton was enormously exhilarated when he met the American mathematician Zerah Colburn in 1820. Colburn was able to perform complex mathematical computations quickly in his head, a skill that enticed the fifteen-year-old Hamilton. The youth had already read Sir Isaac Newton's *Philosophiae Naturalis Principia Mathematica* (1687) and Alexis-Claude Clairaut's *Elémens d'algèbre* (1746) by the time he met Colburn. The excitement generated by his meeting with Colburn led Hamilton in the following year to study the completed volumes of Pierre-Simon Laplace's five-volume *Traité de mécanique céleste* (1798-1825; *A Treatise of Celestial Mechanics*, 1829-1839).

Hamilton's detection of a flaw in Laplace's reasoning brought him to the attention of John Brinkley, a distinguished professor of astronomy at Trinity College who was then also president of the Royal Irish Academy. The following year, when he was seventeen, Hamilton sent a paper he had written on optics to Brinkley, who, upon reading the paper, declared to the Royal Academy that Hamilton was already the most important mathematician of his time.

Hamilton entered Trinity College in 1823. By 1825, he had completed his paper "On Caustics" and submitted it to the Royal Academy, only to be rebuffed because the members of the Academy could not follow his often convoluted reasoning. Hamilton was awarded the *optime* in both classics and mathematics, the first Trinity College student to achieve this dual honor. While still an undergraduate, in 1827, he submitted his paper "Theory of Systems of Rays" to the Royal Academy, establishing with that paper a uniform method of solving all problems in the field of geometrical optics. The paper was of sufficient significance that before he had finished his undergraduate studies at Trinity College, the school's faculty elected William Rowan Hamilton to the Andrews professorship in astronomy, a post that established him as royal astronomer of Ireland and an examiner of graduate students in mathematics at Trinity College. He assumed that post immediately upon graduation.

Life's Work

The post to which the Trinity College faculty elected Hamilton carried with it a residence at the Dunsink Observatory, some five miles from Trinity College. In October, 1827, Hamilton moved into that residence and remained there for the rest of his life. Although he did not have a distinguished career as an astronomer, Hamilton had a large following of people who attended his lectures on astronomy because the range of his literary as well as his mathematical knowledge was sufficient to enliven his presentations.

Hamilton read encyclopedically and regularly wrote poetry, although his friend, the poet William Wordsworth, advised him that his lasting contributions would lie in mathematics rather than in poetry. In 1832, Hamilton published an important supplement to his paper on the theory of rays. This supplement was purely speculative, postulating a new theory about the refraction of light by biaxial crystals. Augustin Fresnel had already developed the theory of double refraction, but Hamilton took the theory an important step beyond where Fresnel had left it. He contended that in certain circumstances, one ray of incident light could be refracted into an infinite number of rays in a biaxial crystal and would be formed in such a way that a cone would then result. Humphrey Lloyd, following Hamilton's speculative lead, proved this theory of conical refraction within two months.

In 1833, after six years of living alone in his official residence, Hamilton—a man of average height and ruddy complexion—married Maria Bayley, whose father had been an Anglican rector in County Tipperary. Maria bore three children, two sons and a daughter. Not renowned for her domestic abilities, Maria presided over a somewhat chaotic household. Hamilton considered liquor a more reliable source of nourishment than anything Maria's cook could provide, and, through the years, he became a heavy drinker.

Hamilton's "On a General Method in Dynamics," published in 1835, brought together his work in optics and dynamics. He proposed a theory that showed the duality that exists between the components of momentum in a dynamic system and the coordinates that determine its position. In many ways, this work was some of Hamilton's most significant, although it took nearly a century for the development of research in quantum mechanics to demonstrate the brilliance and importance of Hamilton's theory.

Hamilton served as the major local organizer of the British Association for the Advancement of Science meeting in Dublin in 1835, an activity that led to his being knighted in the closing ceremonies of that event. In 1837, he ascended to the presidency of the Royal Irish Academy. In 1843, the Crown awarded him an annual life pension of two hundred pounds. During his final illness, Hamilton received word that he had been ranked first on the list of foreign associates of the National Academy of the United States.

The contribution for which Hamilton is best remembered is his discovery of quaternions. This discovery has fundamentally changed the way in which mathematicians deal with three-dimensional space. Hamilton had begun his extensive investigation into ordered paired numbers more than ten years before he made his monumental discovery of quaternions on October 16, 1843, when, during a walk along Dublin's Royal Canal, the answer to a question that had been haunting him for nearly a decade flashed almost supernaturally into his mind. So excited was he by this flash of insight that he carved the formula for his discovery, $i^2 = j^2 = k^2 = ijk = -1$, into the Brougham Bridge.

Hamilton suddenly realized that in three-dimensional space, geometrical operations require not triplets, expressed as i, j, and k and representing space, as had been previously supposed, but rather that, because in three-dimensional space the orientation of the plane is variable, another element, a real term that represents time, must also be considered, resulting in quadruplets rather than triplets. One of the major consequences of this insight was its negation of the previously accepted commutative law of multiplication, which postulates $(a \times b) = (b \times a)$.

Hamilton's work with quaternions, to which he devoted the last two decades of his life, was essential to the development of vector analysis. More recently, further important applications of his theory of quaternions have been instrumental in the description of elementary particles. Hamilton published his *Lectures on Quaternions* in 1853, and his influential *The Elements of Quaternions* appeared posthumously in 1866. William Rowan Hamilton died of gout on September 2, 1865, after a lingering illness.

Significance

Sir William Rowan Hamilton's name lives in both the history of mathematics and the histories of physics and optics. His pioneering work in vector analysis forced specialists in that field to abandon the theory of double refraction and to replace it with Hamilton's expanded theory of conical refraction. The work that led to these changes began while Hamilton was still an undergraduate at Trinity College and reached its culmination in the supplement to his "Theory of Systems of Rays" in 1832.

Hamilton's next significant achievement posited a duality between the components of momentum in a dynamic system and the coordinates that determine its position, a theory that reduces the field of dynamics to a problem in the calculus of variations. This theory came to have considerable significance as the field of quantum mechanics developed.

Hamilton's most memorable contribution by far, however, was his discovery of quaternions, which forced mathematicians to break with the commutative law of multiplication. In its simplified form, termed vector analysis and adapted by J. Willard Gibbs from Hamilton's theory, Hamilton's theory of quaternions has been of great significance to modern mathematical physicists.

R. Baird Shuman

Further Reading

Bell, Eric Temple. *Development of Mathematics*. New York: McGraw-Hill, 1940. Bell relates Hamilton to some of the salient mathematical developments of his time. The coverage is sketchy and has been superseded by Thomas L. Hankins's biography (see below).

_______. *Men of Mathematics*. New York: Simon & Schuster, 1965. Bell puts Hamilton in historical perspective. The chapter "An Irish Tragedy" focuses on

Hamilton, but, although interesting, it is not factually dependable in all respects.

Crilly, A. J. *Arthur Cayley: Mathematician Laureate of the Victorian Age*. Baltimore: Johns Hopkins University Press, 2005. Cayley (1821-1895) was a contemporary of Hamilton; the two men devised a matrix algebra theory that bears their names. Although focusing on Cayley, this biography also describes Hamilton and others who were part of a nineteenth century British mathematical vanguard.

Graves, R. P. *Life of Sir William Rowan Hamilton*. 3 vols. London: Longmans, Green, 1882. The three enormous volumes of this set include extensive selections from Hamilton's correspondence, poetry, and miscellaneous writings, as well as extensive commentary. The work, remarkable in its time for its thoroughness, is badly dated and suffers from lack of selectivity.

Hamilton, William Rowan. *The Mathematical Papers of Sir William Rowan Hamilton*. 4 vols. Cambridge, England: Cambridge University Press, 1931-2000. Volume 1, *Geometrical Optics* (1931), and volume 2, *Dynamics* (1940), are edited by A. W. Conway and J. L. Synge; volume 3, *Algebra* (1967), is edited by H. Halberstam and R. E. Ingram. Volume 4 (2000) is edited by Brendan Scaife and includes Hamilton's *Systems of Rays*, two lengthy letters regarding definite integrals and anharmonic coordinates, and reprints of numerous papers about geometry, astronomy, and other topics. Volumes 1 and 3 contain useful introductions. Despite some omissions, these volumes are superbly produced, and the highest standards of scholarship have been observed in their editing.

Hankins, Thomas L. *Sir William Rowan Hamilton*. Baltimore: Johns Hopkins University Press, 1980. Hankins's critical biography of Hamilton is the definitive work in the field. Meticulously documented, the book is written in such a lively style that it at times reads like a novel rather than like the eminently scholarly work that it is. The best book to date on Hamilton.

James, Ioan. *Remarkable Mathematicians: From Euler to von Neumann*. New York: Cambridge University Press, 2002. This collection of brief biographies of prominent mathematicians includes a seven-page biography of Hamilton.

Synge, J. L. *Geometrical Optics: An Introduction to Hamilton's Method*. Cambridge, England: Cambridge University Press, 1937. Highly technical in nature, this book contains a brief but valuable preface. This book is for the specialist rather than the beginner.

STEPHEN HAWKING

English physicist and cosmologist

Many consider Hawking to be the greatest physicist of the late twentieth century. His work combined the two primary developments of early twentieth century physics—general relativity and quantum mechanics—to explain the origins and structure of the universe. His 1988 book A Brief History of Time was a best seller that introduced studies of the universe to nonscientists.

Born: January 8, 1942; Oxford, England
Also known as: Stephen William Hawking (full name)
Areas of achievement: Physics, astronomy, mathematics

Early Life

Stephen Hawking was born on the three hundredth anniversary of the death of one of the greatest physicists of all time, Galileo Galilei, who is generally credited with proving that Earth revolves around the Sun. Hawking's birth also came only four days after the three hundredth birthday of another great physicist, Isaac Newton, who developed a mathematical model to explain the structure of the universe that was essentially unchallenged for over two hundred years. Hawking was born in Oxford, England, and both of his parents had attended Oxford University. However, the Hawkings had only recently returned to Oxford to escape the likelihood of London being bombed during World War II.

Hawking's father, Frank, was a physician who had the same ambition for his son. After the war, the family returned to London, where Frank directed the parasitological division of the National Institute for Medical Research. The family moved again in 1950 to St. Albans in Hertfordshire, where, at age eight, Hawking enrolled in St. Albans School. He was an unexceptional student there, although his mathematics teacher proved to be an early inspiration to him.

Despite his father's encouragements, Hawking did not find medicine and biology theoretically rich enough and instead decided to major in physics at University College, Oxford. By his own admission, Hawking averaged barely one hour per day of studying at Oxford and decided to concentrate on theoretical physics as a way to avoid the busywork of memorizing facts. After graduating from Oxford, Hawking had hoped to study at Cambridge University with Fred Hoyle, who had developed steady-state cosmology, which argues that the structure of the universe remains relatively constant over time. However, Hawking's acceptance to Cambridge was contingent on his receiving honors from Oxford. Because of his lack of studying, his final examination scores at Oxford were only borderline for an honors degree. In an interview, he then told his examiners that if they gave him honors, he would go to Cambridge. Otherwise, he would stay at Oxford. They awarded him a first-class honors degree in natural science in 1962.

Upon reaching Trinity Hall, Cambridge, Hawking was disappointed to learn that he would not study with Hoyle but with Dennis Sciama, another steady-state cosmologist, but one unfamiliar to Hawking. However, Sciama turned out to be much more available and open to students developing their own alternative perspectives than Hoyle would have been. Within a few months of arriving at Cambridge, Hawking faced a far more serious disappointment. Although he was coxswain on a university crew team, he was never robust or athletic, and he was becoming increasingly clumsy. At his parents' insistence, he saw a doctor. He was diagnosed with amyotrophic lateral sclerosis, or ALS (known as Lou Gehrig's disease in the United States), a degenerative condition causing the gradual loss of control of all muscles, including those necessary to move, gesture, speak, and swallow. Hawking was told he would live perhaps another two years and entered a deep depression as he contemplated the futility of trying to complete his doctorate. Sciama persuaded him to continue but refused to lower his standards.

In 1965 a turning point in Hawking's outlook for the future came when he married Jane Wilde, also a university student. As his condition worsened, he came to need a cane, then a wheelchair, and then an artificial speech synthesizer. However, he overcame the depression and survived despite the diagnosis. After completing his doctorate, he became a research fellow at Cambridge.

Through Sciama, Hawking met Roger Penrose, a mathematician who had developed the idea of a "singularity," a point at which the laws of mathematics and science break down. Hawking earned his doctorate by proposing that singularities could be used to understand the structure of the universe. He came to rethink the concept of singularity and argued that the laws of physics are continuous throughout the universe.

Life's Work

Early in the twentieth century, Newton's physics faced two challenges: relativity, conceived by Albert Einstein, and quantum mechanics, which had several founders. Although Einstein contributed to the development of quantum mechanics, he was uncomfortable with it. What most disturbed him was the uncertainty principle of Werner Heisenberg, which suggests that not everything is knowable and measurable. Because Einstein's relativity insists that everything can ultimately be determined, a position that it shared with Newtonian mechanics, Hawking calls it a "classical theory." One implication of relativity that Einstein himself shunned was that large stars could collapse into black holes, single points with such overwhelming gravitational strength that nothing, including light, can escape. Penrose considered black holes to be singularities and thought that since they emitted no signals, they were unknowable by the laws of physics as scientists understood them.

In the 1920's, astronomers began to believe that the universe was expanding. Hoyle's steady-state cosmology was one model of an expanding universe. However, a rival cosmology emerged to account for the expanding universe, which Hoyle disparagingly dismissed as the "big bang." The big bang theory proposes that billions of years ago the entire universe was compressed into a single point that exploded. The energy released from that explosion produced the four forces that govern the universe (gravity, electromagnetism, strong nuclear force, and weak nuclear force), as well as elementary particles, atoms, galaxies, stars, and planets. The initial explosion was so strong that it continues to cause the universe to expand to the present day. The big bang hypothesis suggests that residual radiation from that explosion should pervade the cosmos even now. Around the time Hawking wrote his doctoral dissertation, the radiation was discovered. This caused most physicists to reject Hoyle's steady state. In his dissertation, Hawking suggested that the entire universe was originally a singularity like a black hole and that the big bang could be understood by comparing the universe to a star collapsing into a black hole. Hawking's analogy reversed the time sequence of a compressing star: The universe explodes from a singularity rather than imploding into one.

One of the most serious problems in twentieth century physics is that its two main theories, relativity and quantum mechanics, appear to be incompatible. Black holes are predicted by relativity, which implies that they should emit no energy. In the 1970's, Hawking began to wonder if he could reconcile relativity and quantum mechanics by applying quantum mechanics to black holes. He found that according to quantum mechanics, black holes would indeed emit energy—a phenomenon since known as Hawking radiation—and would eventually explode. Hence, if the universe prior to the big bang was analogous to a black hole, it would one day burst like a black hole; therefore, the big bang could be explained by combining relativity and quantum mechanics. This means that the primordial universe as well as black holes can be understood by laws of mathematics and science and neither are really singularities. In a 2005 journal article, Hawking modified his theory of black holes to allow that information is never truly lost inside them and is eventually released into the universe again, if in garbled form.

Hawking came to believe that the universe was continuous and governed by a single set of laws. This would not mean that all events were predictable because a comprehensive theory of the universe would still contain Heisenberg's uncertainty principle. Hawking cautions that even if humans knew all the laws that underlie the operation of the universe, an account of all possible occurrences would require knowing the history of every particle, something clearly impossible within finite time.

Modern physics believes that the universe is governed by the four forces of gravity (which binds the planets, stars, and galaxies together), electromagnetism (which binds the atom together), the strong nuclear force (which holds the atomic nucleus together), and the weak nuclear force (which is necessary to account for radioactive decay). Einstein spent his later years unsuccessfully searching for a "grand unified theory," a single set of equations that would account for all four forces. Hawking was convinced that Einstein failed because he did not incorporate quantum mechanics. According to Hawking, the place in time to search for the grand unified theory is the time of the big bang, when, according to Hawking and other physicists, the four forces were one.

The universe assumed the shape it did because of the particular way energy happened to have been distributed at the time of the big bang. With slight differences, the galaxies and stars may never have developed. Indeed, the big bang may have produced regions where space, matter, and energy assumed different forms. Hawking proposed that the result could be an infinite number of "baby universes" and that the universe in which humans live may be only one of many possibilities.

The ultimate fate of the universe may depend upon how much matter it contains. If it contains only the matter that astronomers can see with visible light, then the universe should expand forever. If there is more matter, the universe will eventually contract and ultimately collapse into a single point. Hawking hypothesizes that black holes may have formed not only out of imploding stars but also from residues of the big bang. If that is true, black holes may pervade the universe, and there may be enough invisible matter to one day reverse its expansion. Long after the universe again congeals into a single point, it will again explode and expand. Hence, the universe would have a continual history whose broad outline is predictable. If Hawking is correct, then the dream of physics, a single set of laws to explain the development of the universe and all its contents, is indeed attainable.

Hawking became the Lucasian Professor of Mathematics at Cambridge in 1979, a professorship once occupied by Newton, and he is a Fellow of Gonville and Caius College. Believing that the public should be kept informed about scientific theory, in 1988 he published *A Brief History of Time*, a popular book about the intellectual history of cosmology. To the surprise of all it remained a best seller for 237 weeks, and it sold more than 9 million copies. Other popular books followed: *Black Holes and Baby Universes* (1993), *The Universe in a Nutshell* (2001), and *A Briefer History of Time* (2005).

In 1985, Hawking had to have an emergency tracheotomy after suffering from pneumonia. He then required round-the-clock nursing care, and he lost his ability to speak. A computer system was built for him that enabled him to speak with an electronic voice—with an American English accent. He could control the system with his little finger or by blinking to activate an infrared sensor. He wrote of his condition,

> I am quite often asked: How do you feel about having ALS? The answer is, not a lot. I try to lead as normal a life as possible, and not think about my condition, or regret the things it prevents me from doing, which are not that many.

Many awards and honors have come to Hawking for his work in science. He was created a Commander of the Order of the British Empire in 1982 and Companion of Honor in 1989. He is a member of the Royal Society, an Honorary Fellow of the Royal Society of Arts, and a member of the National Academy of Sciences in the United States. He received the Eddington Medal (1975), the Hughes (1976) and Copley medals (2006) of the Royal Society, the Albert Einstein Medal (1979), the Wolf Prize in Physics (1988), Prince of Asturias Awards in Concord (1989), the Julius Edgar Lilienfeld Prize of the American Physical Society (1999), the Michelson Morley Award (2003), and the Smithson Bicentennial Medal (2005).

Hawking also was a popular public speaker. In addition to physics, he often discussed his worries about the dangers to humanity's future from disease, nuclear weapons, and global warming. In 2006 he feared that global warming might already have proceeded too far to reverse and that humanity's chance for long-term survival might require colonizing other planets. Hawking considered his own work incomplete. However, as he suggested, science is never complete: It is supposed to be self-critical and forever subject to revision.

Stephen and Jane Hawking divorced in 1991. In 1995 he married Elaine Mason, a nurse. He had three children and one grandchild. To celebrate his sixty-fifth birthday in 2007, he took a trip on a specially designed plane that allowed him to float in zero-gravity, the first paraplegic to do so.

Significance

Hawking's theories led to a reassessment of Einstein's reworking of Newton's theory. Newton's theory had served as the foundation of physics for more than two centuries. Hawking, in turn, would not have developed his model had Einstein not preceded him, nor would Einstein have produced relativity without Newton's prior framework. The ideal for scientists is to build on each other's work. Modern scientific truths could become the scientific "errors" of the future, but these truths (and errors) nevertheless are foundational for future work in the sciences. As Newton famously admitted, "I can see far because I stand on the shoulders of giants." Hawking has proved to be another "layer" in a pyramid of giants.

There are critics who charge that Hawking simply was one of a number of scientists who tried to unify physics and to explain the development of the universe, and that his disability is the focus of the media. Even if true, the claim cannot take away Hawking's actual accomplishments in physics and astronomy while living with Lou Gehrig's disease. Einstein, Bertrand Russell (who revised mathematics and worked for world peace), and Linus Pauling (who won Nobel Prizes in chemistry and peace) are remembered not only as great scientists but also as great humanitarians. Hawking, too, has been a great scientist and a great humanitarian. His life story inspires both the disabled and the able-bodied.

Yale R. Magrass

Further Reading

Benford, Gregory. "Leaping the Abyss." *Reason* 4 (April, 2002): 24-31. An entertaining article by a fellow physicist and a writer of science fiction. Benford reports on a conversation with Hawking in which the two discuss the latest cosmological theories and their philosophical underpinnings. Much is revealed about Hawking's character and manner of thinking.

Hawking, Jane. *Music to Move the Stars: A Life with Stephen Hawking*. New York: Macmillan, 1999. Hawking's first wife frankly relates the story of her life with Hawking. She tells of how she suppressed her own personality and ambitions to care for him and that she had problems with his nursing staff, which led to the breakup of the marriage. Includes family photographs.

Hawking, Stephen. *A Brief History of Time*. New York: Bantam Books, 1988. Hawking's best-selling book, written for general readers. Discusses his own work as part of the development of twentieth century physics.

_______. *A Briefer History of Time*. Rev. ed. New York: Bantam Books, 2005. This is a shorter, simpler version of Hawking's 1988 book. Includes discussion of new developments in physics, such as string theory.

_______. *Black Holes and Baby Universes*. New York: Bantam Books, 1993. A collection of popular essays written by Hawking. Examines his life, his philosophy of science, the possibility of science knowing all there is to know, and the origin, structure, and fate of the universe.

_______. *The Universe in a Nutshell*. New York: Bantam Books, 2001. Accompanied by beautiful color illustrations, this work examines cosmological theories, including that of a cosmos with multiple universes. For general readers with some familiarity with physics and astronomy.

Hawking, Stephen, and Roger Penrose. *The Nature of Space and Time*. Princeton, N.J.: Princeton University Press, 1996. This is the published edition of a series of debates between Hawking and his former mentor and collaborator, Roger Penrose. They contrast their opinions about black holes, the origin and fate of the universe, and the philosophy of science.

McEvoy, J. P., and Oscar Zarate. *Introducing Stephen Hawking*. New York: Totem, 1997. A comic-book presentation of Hawking's life and work that is particularly good at mixing graphics and text. In the process of explaining Hawking's work, it offers an amusing but informative history of twentieth century physics.

Singh, Simon. *Big Bang: The Origin of the Universe*. New York: Fourth Estate, 2004. This history of the development of the big bang theory includes discussion of the contributions of many physicists and cosmologists.

White, Michael, and John Gribbin. *Stephen Hawking: A Life in Science*. New York: Penguin, 1993. A biography of Hawking that shows how his personal life history impacted his scientific discoveries.

CAROLINE LUCRETIA HERSCHEL

German-born English astronomer

Herschel was an astronomer and mathematician who spent years assisting her better-known brother William Herschel in his astronomical observations. However, she also independently scanned the sky for new objects, discovering two galaxies, several open star clusters, and eight comets. After her brother's death, she completed the Herschel catalog of twenty-five hundred nebulae, which was included in her nephew John Herschel's General Catalogue.

Caroline Herschel, at age ninety-two. (Library of Congress)

Born: March 16, 1750; Hanover (now in Germany)
Died: January 9, 1848; Hanover
Areas of achievement: Astronomy, music, mathematics

Early Life

Caroline Lucretia Herschel (KEHR-eh-lihn lew-KREE-shee-ah HUR-shehl) was the younger of two girls born to Isaac and Anna Ilse Herschel. She also had four brothers as well as three other siblings who died. Isaac, a musician and professor of music, encouraged his children to become educated. As the youngest daughter, Caroline was expected by her mother to do most of the housework, who generally treated Caroline as she would a servant.

When Caroline was seven, her brothers William and Alexander went to England when their regimental band was ordered there and her father fought in the war against France. At age ten, Caroline contracted typhoid fever, which stunted her growth and caused her to stoop slightly. She was only four feet, three inches tall as an adult, and her father predicted that she would likely never marry. In spite of her mother's attempts to enslave Caroline to domestic duties, her father educated her in music and mathematics and introduced her to astronomy. By 1760 her father had returned from the war, but he remained in poor health until his death in 1767.

In spite of her mother's attempts to keep her in servitude and after her father's death, Caroline decided to take more control of her own life and so learned dressmaking, planning to become a governess. Her brother William had been hired as organist in Bath in 1766, and by 1772 he decided to invite Caroline, now twenty-two years old, to live with him. Against her mother's protests she moved to England, where her brother gave her voice lessons and tutored her in English and mathematics.

She soon began a singing career under her brother's direction. She was a first soprano singing George Frideric Handel's *Messiah* and *Judas Macabaeus*, among other works, and was often engaged several nights a week. She turned down offers from other directors, reserving her singing for performances directed by her brother. During this time she also began helping William with his telescope projects. She learned to grind and polish lenses and construct the finished telescopes.

Life's Work

As she learned the skills needed to help William in the shop, Caroline also continued her studies in mathematics, focusing almost exclusively on topics applicable to astronomy, her brother's growing passion. William's obsession with his telescopes became so consuming that Caroline sometimes had to feed him while he continued his work, and other times she would read to him to pass the time productively as he laboriously turned yet another lens for a new telescope.

By 1781, William was spending more and more time gazing through his telescopes at the night sky, and Caroline was often by his side taking notes on his observations. In this same year, William made his breakthrough discovery, the planet Uranus. The astronomical community praised his discovery and King George III awarded William an annual salary of £200. William resigned his position as organist, and soon thereafter Caroline gave up her music career to devote her time to astronomy as well.

As William built ever-larger telescopes and needed increased workspace, he and Caroline moved to larger quarters, first to Clay Hall and then to Slough. Although Caroline spent most of her time assisting her brother by taking records during their nightly observations and doing calculations during the day so the night's objects could be cataloged, she was still able to find time for her own observations. William encouraged her in this direction. She discovered her first new object on February 26, 1783, an open star cluster in the constellation Canis Major. Over the next several years she discovered more than ten other objects, including a large spiral galaxy and another smaller galaxy.

Caroline also spent a considerable amount of effort sweeping the sky methodically for comets, a common pursuit of amateur and professional astronomers alike. On August 1, 1786, she discovered her first comet, the first comet discovered by a woman. Between 1786 and 1797 she discovered seven more comets. Six of the eight she found in total bear her name. In recognition of her continuing work with her brother, King George III granted her an annual salary of £50, £150 less than William. When William married in 1788, Caroline lamented the change in her relationship with her brother. She moved to private lodgings nearby but continued to work with William and in time got along well with his wife, at which point she moved back into William's home.

After 1797, Caroline devoted most of her effort to other pursuits. In 1798 she published a work that cross-referenced and corrected John Flamsteed's *Historia coelestis Britannica* (1725). In addition to revising Flamsteed's well-known catalog, she added an additional 560 stars to the list. Over the next twenty-five years, until her brother's death in 1822, Caroline devoted most of her time to helping educate her brother's son, John, who was born in 1792.

When her brother died she returned to Hanover, resumed her work, and produced the Herschel catalog, listing twenty-five hundred nebulae, many of them discovered by William. The catalog was completed and then published as *General Catalogue of Nebulae and Clusters of Stars* in 1864 by nephew John.

For her foundational work she received the gold medal in science 1828 from the Royal Astronomical Society. Additional honors followed in 1835, when she was elected an honorary member of the Royal Society, and in 1838, when she was elected to the Royal Irish Academy. In 1846 she received a gold medal in science from the king of Prussia.

During the remaining two years of her life her health gradually worsened. She died on January 9, 1848. Her epitaph, which she composed herself, reads, "The eyes of her who is glorified here below turned to the starry heavens."

Significance

Caroline Lucretia Herschel's fame and significance certainly were muted after living and working in William's more famous shadow. Although she made important astronomical discoveries of her own, those discoveries are not considered comparable in significance to those of William. Separating William's discoveries from her own, though, may be unfair. The two worked side by side for years, sharing in their work. Without her devotion to William's work, it is possible that his accomplishments would have been minimized. Certainly he could have worked with a different assistant, but Caroline's skills were so extraordinary that she could be considered more a colleague than an assistant. She certainly would have had greater opportunities for education if she were born male. Nevertheless, her perseverance enabled her to surmount many of the usual barriers to women's intellectual advancement. Her brother also helped remove her from the domestic life in which her mother attempted to trap her. Her illness as a child left her disfigured and ruined her chances at marriage, but it also allowed her the time to pursue a career.

Discovering several new objects in the sky and eight comets would have assured any astronomer some degree of fame. What makes Caroline's work more impressive is that she received recognition as a woman. Caroline Lucretia Herschel represents the first well-known woman astronomer, breaking ground for other women to follow. Her life was commemorated in 1889 with an asteroid bearing her middle name, Lucretia. In 1935 a lunar crater was also named for her.

Bryan D. Ness

Further Reading

Herschel, Caroline Lucretia. *Caroline Herschel's Autobiographies*. Cambridge, England: Science History, 2003. Contains annotated editions of the two autobiographies penned by Herschel.

Higgins, Frances Lowry. *Sweeper of the Skies: A Story of the Life of Caroline Herschel, Astronomer*. Chicago: Follett, 1967. A biography of Herschel focusing primarily on her astronomical work.

Hoskin, Michael A. *The Herschel Partnership: As Viewed by Caroline*. Cambridge, England: Science History, 2003. A biographical work that focuses on the partnership between Caroline and William.

Lubbock, Constance A. *The Herschel Chronicle: The Life-Story of William Herschel and His Sister, Caroline Herschel*. Cambridge, England: Cambridge University Press, 1933. A biography of William and Caroline with material drawn from their own records.

Wearner, Robert G. *Caroline Herschel: First Woman Astronomer*. Austin: University of Texas Press, 1989. A biography focusing primarily on Herschel's astronomical work.

WILLIAM HERSCHEL

German-born English astronomer

Herschel discovered Uranus and six of its moons. Using several large telescopes that he designed and manufactured, he made numerous planetary observations and also discovered that the Milky Way has a cloud-like appearance because of its numerous distant stars. He also identified double and triple stars, nebulae, star clusters, and other celestial objects, and made detailed studies of the structure of comets and the surface of the Sun.

Born: November 15, 1738; Hanover (now in Germany)
Died: August 25, 1822; Slough, Buckinghamshire, England
Also known as: Friedrich Wilhelm Herschel (birth name)
Areas of achievement: Astronomy, science and technology, mathematics, music

Early Life

William Herschel was one of six surviving children of Isaac and Anna Ilse Herschel. Isaac was a skilled musician and professor of music, and his talents seemed to have passed to several of his children, including William, who studied oboe, violin, and organ. William also loved mathematics and gained an interest in astronomy from his father.

In 1755, William and his brother Jacob, members of the Hanoverian regimental band, traveled to England when their regiment was ordered there. Initially, William's pay was barely enough to avoid poverty, but as his musical abilities gained recognition he was invited to be bandmaster in a militia regiment in northern England. He forged relationships with wealthy families in the area, which led to even more musical involvement, including teaching instrument lessons, and to better fortunes.

In 1764, Herschel paid a brief visit to his family in Germany. Upon his return to England in 1765 he was appointed organist at Halifax, further securing his economic fortunes. Within a year he moved up to the more lucrative position of organist at the Octagon Chapel in Bath. His sister Caroline Lucretia Herschel joined him in 1772. By this time William Herschel's life was extremely busy. What most would have considered a heavy, full-time job as a talented and popular musician was supplemented in Herschel's case with his pursuit of mathematics.

Life's Work

Soon after receiving a small telescope, William Herschel began the work that would gradually consume him for the rest of his life. His new thirst for exploring the night sky led him to build a larger telescope, since ready-made models were too expensive. After studying optics and experimenting with lens and mirror production, he settled on making an 18- to 20-foot telescope.

Soon Herschel's obsession consumed not only much of his free time but also several rooms of the house that were converted into workshops. Caroline, along with brother Alex, were also drafted to work on his projects. All of this activity was taking place amid Herschel's numerous musical responsibilities. His fixation on completing the 20-foot telescope was so intense that he would not stop turning a lens to eat, and Caroline often fed him or read to him while he worked. However, she was not simply a secretary and helpmate; she was also was an astronomer in her own right, and she possessed many of the observational and mathematical skills of her brother.

On the night of March 13, 1781, Herschel discovered the planet Uranus, initially named Georgium Sidus in honor of King George III. Consequently, he was awarded the Copley Medal and was elected a fellow in the Royal

Society. The following year George III appointed him King's Astronomer and awarded him an annual pension of 300 guineas and a home nearer the king. With a steady, dependable income, Herschel finally made a permanent switch to astronomy, resigning his post as organist at Bath.

Herschel's next major project was to construct a more powerful 40-foot telescope. He eventually received funding of £2,000, and, by 1786, he and his sister had moved to Slough, where he built a better observatory and began constructing a larger telescope. Because of the size and great weight of this larger telescope, everything about it had to be specially designed. The end product was so cumbersome that Herschel needed assistance to adjust it for viewing specific portions of the sky. Caroline continued to assist him, devoting her life to his work. She was his primary record keeper, spending most nights by his side at the telescope. Because of Herschel's growing popularity, King George also provided funds to hire work assistants, who also gave tours to the growing numbers of visitors who came to see the large telescope and observatory.

Herschel's primary interest was in the planets, but he also made observations of the Sun and developed theories about its surface appearance. He performed experiments using a prism to separate sunlight into its component colors, measuring the temperature of each. As a result, he discovered infrared radiation—invisible rays beyond the red end of the spectrum—and showed they possessed heat.

He observed the surface of the Moon in great detail and attempted to measure the height of some of its surface features. He believed that some of the Moon's bright spots were volcanoes, a theory now known to be incorrect. He determined the shapes and approximate sizes of several of the "minor" planets, and believed that they were too small to be planets. He coined the term "asteroids" for these bodies, a name still used. He made detailed observations of comets, noting especially their nuclei and the apparent production of their own light.

In 1788 he married a woman with a large inheritance, which enabled him to devote his time completely to astronomy. His new marriage briefly strained his relationship with his sister. She moved out for a time but later rejoined the household and continued her work with Herschel.

Herschel was awarded an honorary degree from Oxford, and in 1816 he was knighted, becoming Sir William Herschel. In 1820 he was elected the first president of the Astronomical Society. By this point Herschel's health was beginning to fail; he died on August 25, 1822. Both Caroline and Herschel's son John carried on his work after his death, preparing a number of his documents for posthumous publication.

Significance

William Herschel has taken his place among the great astronomers primarily because of his discovery of Uranus and six of its moons. His meticulous nature, both as an observer and as a telescope-builder, helped him in his discoveries. His lesser-known, though no less important, discoveries stem from these same, almost obsessive, tendencies. It is fair to say that his successes also lay in large part with his sister, Caroline. Her devotion to his work, along with her own abilities in astronomy and mathematics, enabled him to accomplish much more than he might have on his own.

Herschel's superior telescopes also helped him see objects never before observed by other astronomers. Most notable was his discovery that many hazy objects in the sky are actually groups of stars, often with clouds of gas accompanying them (nebulae). He discovered many nebulae and showed that the Milky Way is composed of millions of stars. Some of his speculations, however, about the Moon, the Sun, and the planets have not withstood the test of time. One of his wildest speculations was that all planets, and even the Sun, were inhabited.

Herschel's work has been commemorated by several objects bearing his name, including a crater on the Moon and one on Mimas, a moon of Saturn; a large impact basin on Mars; and an asteroid. The William Herschel telescope, located in La Palma on the Canary Islands, also bears his name. In 2007 the European Space Agency plans to launch the Herschel Space Observatory, which will be the largest space telescope of its kind.

Bryan D. Ness

Further Reading

Armitage, Angus. *William Herschel.* Garden City, N.Y.: Doubleday, 1963. A biography of William Herschel focusing on his musical as well as his astronomical work.

Dreyer, John Luis Emil, ed. *The Scientific Papers of Sir William Herschel.* Bristol, England: Thoemmes Continuum, 2003. A collection of previously unpublished papers edited under the direction of a joint committee of the Royal Society and the Royal Astronomical Society. Includes a brief introductory biography.

Hoskin, Michael A. *William Herschel and the Construction of the Heavens*. London: Oldbourne, 1963. A biography of William Herschel focusing primarily on his astronomical work.

Lubbock, Constance A. *The Herschel Chronicle: The Life-Story of William Herschel and His Sister, Caroline Herschel*. Cambridge, England: Cambridge University Press, 1933. A more personal biography of Herschel and his sister, with material drawn from their respective records.

JOHANNES AND ELISABETHA HEVELIUS

German-Polish astronomers

Johannes and Elisabetha Hevelius compiled the most accurate nontelescopic catalog of the stars. Johannes established an observatory in Gdansk, compiled detailed maps of the Moon, and published a book describing many comets, including four that he discovered. Elisabetha published a collection of engravings of constellations, including eleven new constellation groupings, seven of which are still in use.

Johannes Hevelius

Born: January 28, 1611; Gdansk, Poland
Died: January 28, 1687; Gdansk
Also known as: Johann Hewel or Howelcke; Jan Heweliusz

Elisabetha Hevelius

Born: c. 1647; Gdansk?
Died: 1693; Gdansk?
Also known as: Catherina Elisabetha Koopman (given name)
Areas of achievement: Astronomy, science and technology, scholarship

Early Lives

Johannes Hevelius (hay-VAY-lee-uhs) was born to a noble family of at least ten children. His father was a prosperous property owner who operated a brewery. His early education was at a gymnasium in Gdansk from 1618 to 1624. He then spent three years at a school near Bromberg, Poland, before returning to the Danzig gymnasium for three more years. There he studied under a tutor who gave him private lessons in astronomy and instrument-making, in addition to the regular curriculum.

Johannes traveled to the University of Leiden in the Netherlands in 1630, where he studied law for one year, plus some mathematics and optics. During the next two years, he visited several scientists in Paris. He then returned to Gdansk to work in his father's brewery and to study the constitution of Gdansk. In 1635, he married Katharina Rebeschke, the daughter of a wealthy Gdansk landowner. After observing the solar eclipse of June 1, 1639, Johannes began a lifetime of astronomical work and a long career of civic responsibilities.

In the year following his first wife's death in 1662, Johannes married Catherina Elisabetha Koopman, the well-educated sixteen-year-old daughter of a wealthy merchant. Although she was thirty-six years younger than her husband, she became his chief assistant in his astronomical work while hosting many visiting astronomers and raising three daughters.

Lives' Work

Johannes devoted his life to civic duties, corresponded with foreign astronomers, and worked on astronomy. His first civic office in Gdansk, begun in 1641, was honorary magistrate; in 1651, he became a city councillor. The astronomers he communicated with included the Dutchman Christiaan Huygens, who discovered the rings of Saturn in 1656; John Flamsteed, who was the first royal astronomer in England; and Edmond Halley, who first demonstrated the elliptical orbits of comets and became the second royal astronomer in England. His major occupation, however, was astronomy.

Starting with a small upper room as an astronomical observatory, Johannes added a small roofed tower in 1644, and later built a platform on his house with both a stationary and a rotating observatory. In 1644, he was the first to observe the phases of Mercury, further confirming the heliocentric system. After his father died in 1649, he took over the brewery and used the extra funds for telescopes and other astronomical equipment, which made his observatory, called Sternenburg, one of the finest in the world. He also manufactured his own instruments in his own workshop and had a printing press for publishing his observations.

In 1647, Johannes published his first important work, an atlas of the Moon called *Selenographia: Sive, lunae descriptio* (reproduced map, with English translation, 1966). In the introduction to this book, he illustrates and describes his optical lathe for grinding telescope lenses. He also describes several telescopes, including the largest one, which was some 11 feet long and had about a 50-power magnification. He includes a drawing of Saturn with two semicircular handles, records of the movements of the moons of Jupiter, and observations of sunspots and eclipses. In the rest of the book, he describes lunar markings, the apparent lunar motions called librations, and drawings in his own hand of the Moon. His engravings of the lunar surface are among the best of the seventeenth century and introduce many of the names of lunar craters, mountains, and other features that are still used today.

Johannes's second great work, *Cometographia* (study of comets), published in 1668, twenty-one years after his first book. Following some introductory engravings, including one of his house and observatories, he turns to discussion of the comet of 1652 and gives evidence contrary to Aristotle that it is located in space beyond the Moon. He also provides extensive information about comets of the two preceding centuries, including four that he discovered, and suggests that they move on parabolic paths around the Sun at the focus.

Johannes described his work in designing and making his astronomical instruments (celestial machines) in a two-volume work called *Machina coelestis* (1679), partially translated as *The Illustrated Account Given by Hevelius in His "Machina celestis" of the Method of Mounting His Telescopes and Erecting an Observatory* (1882). The two volumes describe more than a dozen large (up to 6 feet) quadrants and sextants made of copper and wood for measuring celestial positions, many of them based on designs of the Danish astronomer Tycho Brahe but with even greater accuracy than those of Brahe. He also described the design of several long telescopes with focal lengths ranging from about 25 to nearly 150 feet and lenses up to 8 inches in diameter, including the difficulties of mounting such large instruments. His longest telescopes turned out to be nearly useless, however.

After several assistants either died or were found to be unsatisfactory, Johannes opened his observatory to his second wife Elisabetha, who became his most accurate and diligent observer. She is shown in two plates of *Machina coelestis* assisting in Johannes's observatory. For about ten years, she helped to run the observatory; in 1679, Johannes's house, observatories, instruments, workshop and most of his papers were destroyed by a fire. By the end of 1681, the observatory was rebuilt, but with fewer and inferior instruments. He labored for five more years in declining health before dying on his seventy-sixth birthday.

Elisabetha Hevelius carried on with the work after Johannes died and completed his most important collection of observations. Three years after his death, she published a catalog of the positions and magnitudes of 1,564 stars, a work called *Prodromus astronomiae* (1690; *The Star Atlas*, 1968), the largest star catalog to that date and the last one compiled without the aid of the telescope. It gave star positions to an accuracy of about one minute of arc, about twice as accurate as Brahe's measurements. Also in 1690, Elisabetha published a volume of fifty-six engravings of constellations. Her *Uranographia* (study of the heavens) included eleven new constellation groupings with seven names that are still used.

Significance

Although Johannes Hevelius was one of the leading astronomers of the mid-seventeenth century and the most accurate of the declining breed of nontelescopic observers, he represents more of a transitional figure in the inevitable trend toward telescopic astronomy. He was challenged to use telescopic sights in place of plain sighting instruments by John Flamsteed and later by Robert Hooke. Hooke wrote a critical work on Hevelius's *Machina coelestis* called *Animadversions on the First Part of the Machina coelestis of the Honourable, Learned, and Deservedly Famous Astronomer Johannes Hevelius* (1674). The resulting controversy led the Royal Society in England to sponsor a 1679 visit to Gdansk by Edmond Halley, who concluded that Johannes could determine stellar positions without a telescope as accurately as Halley could with one at that point in time. Seven years later, in private correspondence, Halley altered this conclusion in favor of telescopic sights.

In his lunar studies, Johannes produced the first true atlas of the lunar surface and the best descriptions of librational cycles up to his time, providing a sound basis for further work in this area. His comet studies were especially useful to Halley in determining the elliptical orbits of comets. His most valuable contribution was his catalog of stars, which was reprinted by Flamsteed in 1725 and used well into the eighteenth century. The completion of this catalog by Elisabetha Hevelius, along with fifteen years of work with her husband before he died, established her as one of the rare women who was in a position to make important contributions to science in the seventeenth century.

Joseph L. Spradley

Further Reading

Bell, Louis. *The Telescope*. New York: McGraw-Hill, 1922. The first chapter of this book on "The Evolution of the Telescope" has several pages on the contributions of Johannes Hevelius to the early development of the telescope.

Field, J. V., and Frank A. J. L. James, eds. *Renaissance and Revolution: Humanists, Scholars, Craftsmen, and Natural Philosophers in Early Modern Europe*. New York: Cambridge University Press, 1993. This collection on issues in the scientific revolution includes an essay entitled "Johannes Hevelius and the Visual Language of Astronomy," by Mary Winkler and Albert van Helden, with several copies of engravings from Johannes's *Selenographia*.

Hevelius, Johannes. *The Star Atlas*. Introduced and edited by V. P. Sheglov. Tashkent, Uzbekistan: FAN Press, 1968. A reproduction of 56 plates of Johannes and Elisabetha's star catalog, compiled in cooperation with the Institute of Astronomy at the Academy of Sciences of the Uzbek, formerly part of the Soviet Union. Includes bibliographic footnotes.

McPike, Eugene Fairfield. *Hevelius, Flamsteed, and Halley: Three Contemporary Astronomers and Their Mutual Relations*. London: Taylor & Francis, 1937. This book reviews the controversy over telescopic versus nontelescopic sighting instruments.

Montgomery, Scott L. *The Moon and the Western Imagination*. Tucson: University of Arizona Press, 1999. This book describes Johannes's studies of the Moon in Chapter 11, "Johannes Hevelius: A Moon of Higher Origins," including several copies of engravings from the *Selenographia*.

HIPPARCHUS

Greek astronomer

Hipparchus was the greatest astronomer of ancient times. He was the founder of trigonometry, which he used to determine the distances from Earth to the moon and sun, and the first to use consistently the idea of latitude and longitude to describe locations on Earth and in the sky.

Born: 190 B.C.E.; Nicaea, Bithynia, Asia Minor (now İznik, Turkey)

Died: After 127 B.C.E.; possibly Rhodes, Greece

Early Life

Very little is known about the life of Hipparchus (hih-PAHR-kuhs). He was born in Nicaea, a Greek-speaking city in Bithynia (modern İznik, Turkey), in the northwestern part of Asia Minor. Calculations in his works are based on the latitude of the city of Rhodes, on the island of the same name, so many historians believe that he spent a major portion of his life there. Rhodes was a merchant center, a convenient port from which to make voyages. At least one of Hipparchus's observations was made in Alexandria, so it seems that he visited and perhaps spent time as a student or research scholar at that great nucleus of scientific inquiry. Because he was intensely interested in geography, it is likely that he traveled to other places in the Mediterranean basin and the Near East. He seems to have been familiar with Babylonian astronomy, including eclipse records, but it is impossible to say how he came to know these.

Life's Work

Most of what is known of Hipparchus comes from the *Mathēmatikē syntaxis* (c. 150 C.E.; *Almagest*, 1948) of Ptolemy, whose work depends to a considerable extent on that of the earlier scientist, and from the *Geōgraphica* (c. 7 B.C.E.; *Geography*, 1917-1933) of Strabo. Of Hipparchus's own writings, only the *Tōn Araton kai Eudoxou* phainomenon exigesis (commentary on the phenomena of Eudoxus and Aratus) survives, in three books. It criticizes the less accurate placement of stars and constellations by two famous predecessors. It is certainly not one of his most important works, but it contains some information on his observations of star positions, which were the basis of his lost star catalog. Other lost works of Hipparchus include *Peri eviausiou megethous* (on the length of the year) and *Peri tes metabaseos tōn tropikon kai isemerinon semeion* (on the displacement of the solstitial and equinoctial points). He is also credited with a trigonometrical table of chords in a circle, a work on gravitational phenomena called *On Bodies Carried down by Their Weight*, an attack on the geographical work of Eratosthenes, a compilation of weather signs, and some aids to computational astrology.

A number of achievements are attributed to Hipparchus by Ptolemy and other ancient writers. A new star appeared in the constellation Scorpio in July, 133 B.C.E. Hipparchus realized that without an accurate star catalog, it was impossible to demonstrate that the star was indeed new, so he set about producing a complete sky map with a table of the positions of the stars, including the angle north or south of the celestial equator (latitude) and the angle east or west of the vernal equinox point (one of the two intersections between the celestial equator and the sun's path, or ecliptic).

In order to do this, he needed a means of measuring celestial angles, which led him to invent many of the sighting instruments, including the diopter and possibly the armillary astrolabe, used by astronomers before the invention of the telescope in the seventeenth century. He also knew how to calibrate water clocks. Hipparchus's star catalog included about 850 stars, along with estimates of their brightness. He divided the stars into six categories, from the brightest to the dimmest, thus originating a system of stellar magnitude. He also made a celestial globe, showing the locations of the fixed stars on its surface.

In comparing his own measurements of positions of stars with those of earlier astronomers, Hipparchus discovered that there had been a systematic shift in the same direction in all of them. He noticed the phenomenon first in the case of the bright star Spica. In 283 B.C.E. Timocharis had observed the star to be eight degrees west of the autumnal equinoctial point, but Hipparchus found the figure to be six degrees. He found a displacement for every other star that he was able to check. These discrepancies, he established, were the result of a shift in the position of the equinoxes—and therefore of the celestial equator and poles. In modern astronomy, this shift is called the precession of the equinoxes and is known to be caused by a slow "wobble" in the orientation of Earth's axis. The spot to which the north pole points in the sky (the north celestial pole) describes a circle in a period of more than twenty-six thousand years. Hipparchus was first to describe and to attempt to measure this phenomenon. He was, however, unable to explain its cause, since he held the geocentric theory, which postulates a motionless Earth at the center of a moving universe.

From the beginning of theoretical astronomy, the geocentric theory had been the accepted one. It assumed that the sun, moon, planets, and stars were carried on vast transparent spheres that revolved at different but constant speeds around Earth. Unfortunately, in order to explain the observed motions of the planets, which vary in speed and sometimes are retrograde relative to the stars, astronomers had to postulate the existence of additional spheres, invisible and bearing no celestial bodies but interconnected with the other spheres and affecting their motions. An Alexandrian astronomer, Aristarchus, had proposed the heliocentric theory, which holds that Earth, with its satellite, the moon, and all the other planets revolve around the central sun. The main appeal of this theory was its simplicity; it required fewer imaginary spheres to make it work.

Hipparchus rejected the heliocentric theory and instead adopted modifications of the geocentric theory to make it accord better with observations, perhaps following Apollonius of Perga. The main feature of the Hipparchan system is the epicycle, a smaller sphere bearing a planet, with its center on the surface of the larger, Earth-centered sphere and revolving at an independent speed. He also postulated eccentrics, that is, that the centers of the celestial spheres do not coincide with the center of Earth. The geocentric system with epicycles is often called "Ptolemaic," as Ptolemy made observations to support the theory developed by Hipparchus. Aristarchus's heliocentric theory is closer to the picture of the solar system provided by modern astronomy.

In developing his astronomical system, Hipparchus observed the period of revolution of the celestial objects that move against the background of the stars. That of the sun, which is the year, he found to be 365 1/4 days, less 1/300 of a day, a figure that was closer to the true one than that of any previous astronomer. He noticed the inequality in the lengths of the seasons, which he correctly attributed to the varying distance between Earth and the sun but incorrectly explained by assuming that the center of the sun's sphere of revolution was eccentric to the center of Earth. These conclusions were, perhaps, a step in the direction of recognizing that the relative motion of the two bodies describes an ellipse. He also achieved a measurement of the length of the lunar month, with an error of less than one second in comparison with the figure now accepted. The Roman scholar Pliny the Elder wrote that Hipparchus countered the popular fear of eclipses by publishing a list that demonstrated their regularity over the preceding six hundred years.

Hipparchus attempted to measure the distances of the moon and sun from Earth by observing eclipses and the phenomenon of parallax (the shift in the apparent position of the moon against the background of the stars under changing conditions). His figure for the distance of the moon (60.5 times the radius of Earth) was reasonably accurate, but his estimate of the sun's distance (2,550 times Earth's radius) was far too small. (The true ratio is about 23,452 to 1.) In fairness to Hipparchus, it should be noted that he regarded his solution to the problem of the sun's distance as open to question.

In order to make the mathematical computations required by these problems, it was necessary for Hipparchus to know the ratios of the sides of a right triangle for the various angles the sides make with the hypotenuse—in other words, the values of trigonometrical functions. He worked out tables of the sine function, thus becoming, in effect, the founder of trigonometry.

Geography also occupied Hipparchus's attention. He began the systematic use of longitude and latitude, which he had also employed in his star catalog, as a means of establishing locations on Earth's surface. Previous geographers show evidence of knowing such a method, but they did not employ it consistently. Hipparchus was able to calculate latitudes of various places on Earth's surface by learning the lengths of the days and nights recorded for different seasons of the year, although the figures given by him were often in error. As the base of longitude, he used the meridian passing through Alexandria. He was especially critical, probably too much so, of the descriptive and mathematical errors in the work of Eratosthenes. He even had some quibbles with the famous measurement of the spherical Earth, which is the latter's most brilliant achievement. It may be Hipparchus, rather than Eratosthenes, who first described climatic zones, bounded by parallels of latitude north and south of the equator.

Significance

Hipparchus was a careful and original astronomer whose discoveries, particularly that of precession, were of the greatest importance in the early history of the science. He was a meticulous observer who produced the first dependable star catalog and who determined the apparent periods of revolution of the moon and the sun with an exactitude never before achieved. As a mathematician, he originated the study of trigonometry, compiling a sine table and using it in an attempt to measure distances in space beyond Earth that was, at least in the case of the moon, successful. Both as astronomer and as geographer, he pioneered the systematic use of the coordinates of latitude and longitude. He devised instruments for use in these observations and measurements.

Unfortunately, almost all Hipparchus's writings have disappeared, so modern assessments of his work must depend on ancient writers who happened to mention him. His influence was important enough to cause several later scientists whose works survive to refer to and summarize him. Most notable among these were Ptolemy and Strabo. It is sometimes hard to tell when these authors, particularly Ptolemy, are following Hipparchus and when they are going beyond him to present their own conclusions. Ptolemy's work became the standard textbook on astronomy until the time of Nicolaus Copernicus in the sixteenth century; thus Hipparchus's name was deservedly remembered. One of Hipparchus's most important mathematical successors was Menelaus of Alexandria (fl. c. 100 C.E.), who developed the study of spherical trigonometry.

J. Donald Hughes

Further Reading

Dicks, D. R. *Early Greek Astronomy to Aristotle*. Ithaca, N.Y.: Cornell University Press, 1985. Hipparchus is not given major treatment, although he does appear as an important figure in the history of astronomy. The discussion of his criticisms of Eudoxus and Aratus is particularly good.

Dreyer, John L. E. *A History of Astronomy from Thales to Kepler*. 2d ed. New York: Dover, 1953. This fine, accessible study places Hipparchus clearly in the context of the development of astronomy. Dreyer differs from common interpretation in crediting Ptolemy, not Hipparchus, with the theory of epicycles.

Heath, Thomas. *A History of Greek Mathematics*. 2 vols. Reprint. New York: Dover, 1981. Includes a section on Hipparchus in the second volume, emphasizing his probable contributions to the origin of trigonometry and establishing his place in the history of mathematics.

Lloyd, G. E. R. *Greek Science After Aristotle*. New York: W. W. Norton, 1973. Rather than giving a separate treatment to the subject, this work discusses the contributions of Hipparchus as they arise in a general study of ancient science from the fourth century B.C.E. to the end of the second century C.E. The attention given to Hipparchus is appropriate and appreciative.

Neugebauer, Otto. *A History of Ancient Mathematical Astronomy*. 3 vols. New York: Springer-Verlag, 1975. This work contains a section on Hipparchus in volume 1, briefly discussing what little is known about his life and chronology and devoting the rest of its space to a careful consideration of his astronomical work. There are some mathematical and astronomical symbols and formulas that the layperson may find difficult.

Ptolemy. *Ptolemy's "Almagest."* Translated by G. J. Toomer. Princeton, N.J.: Princeton University Press, 1998. Much of what is known about Hipparchus is based on Ptolemy's words. This fine translation has complete notes and a useful bibliography.

JEREMIAH HORROCKS

English astronomer

Horrocks's observation of the transit of Venus, in 1639, is the earliest on record. He applied Kepler's laws of planetary motion to the Moon, comets, and planets.

Born: c. 1619; Toxteth Park, England
Died: January 3, 1641; Toxteth Park
Also known as: Jeremiah Horrox
Areas of achievement: Astronomy, science and technology

Early Life

Jeremiah Horrocks (jur-uh-MI-uh HAWR-uhks) was born in Toxteth Park, near Liverpool, England, around 1619, but the records from Toxteth Park for that year have been lost. There has been some speculation that his father may have been William Herrocks, a farmer, but historical research suggests that his father was actually James Horrocks, a watchmaker. James Horrocks was married to Mary Aspinwall, who was from a family of influence in Toxteth Park. Horrocks entered Emmanuel College, Cambridge University, as a sizar, or poor scholar, in 1632. At that time, the curriculum included mostly the arts, divinity, and classical languages, although it is likely Horrocks also learned geometry and some classical astronomy. He studied at Cambridge until 1635, when he returned to Toxteth Park, where he is believed to have become a tutor.

In Horrocks's time, ideas about planetary motion were in transition. Since the ancient Greeks, it had generally been accepted that the earth was the center of the universe. In 1543, however, Nicolaus Copernicus published *De revolutionibus orbium coelestium* (1543; *On the Revolutions of the Heavenly Spheres*, 1952; better known as *De revolutionibus*), which argued that it was the Sun that was at the center, with the heavenly bodies, including the earth, moving around the Sun. In 1608, Johannes Kepler's analysis of careful position measurements of the planets and the stars, taken by the Danish astronomer Tycho Brahe, demonstrated that the planet Mars moved in an elliptical path around the Sun, and Kepler developed a series of laws that described planetary motion.

The precision of Kepler's laws of planetary motion may have appealed to Horrocks's uncle, who is believed to have been a clockmaker. Tradition says it was this uncle who first stirred Horrocks's interest in astronomy. Shortly after returning to Toxteth Park, Horrocks constructed his own telescope, and he developed projection techniques that enabled him to view the Sun. He used his telescope to measure the positions of the planets and used these measurements to correct errors in existing astronomical tables.

Life's Work

As Horrocks's interest in astronomy deepened, he constructed simple instruments to measure the size of the Moon, Venus, and the Sun. To measure the size of Venus, he made a pinhole in a piece of paper and mounted the paper on the end of a stick. He determined the size of the pinhole by measuring the size of the pin, using an ingenious technique. Horrocks wrapped thread around the pin many times, then unwound the thread and measured its length. By dividing this length by the number of times he had wrapped the thread around the pin, he was able to accurately determine the circumference of the pin. Then he looked at Venus through the pinhole, adjusting the distance of the pinhole from his eye until Venus just filled the hole. By measuring the distance of the card from his eye, he calculated the angular size of Venus.

The passage of a planet across the disk of the Sun is called a transit, and the prediction of the exact date and time when a transit will occur requires a very precise calculation of the motion of the planet. Only transits of the two inner planets, Mercury and Venus, can be observed from the earth. These transits occur much less frequently than eclipses of the Sun by the Moon. There are generally about thirteen transits of Mercury each century. However, because Venus's orbit is considerably larger than Mercury's orbit, transits of Venus are far rarer, occurring in pairs separated by eight years. More than a century elapses between each transit pair. Seven transits of Venus have occurred since the invention of the telescope (1631, 1639, 1761, 1769, 1874, 1882, and 2004).

The French astronomer Pierre Gassendi was the first person to record that he had observed a planetary transit, that of the planet Mercury in 1631, but the time of this transit differed by six hours from that predicted by Kepler. A transit of Venus occurred just one month later. Gassendi attempted to observe it, but he failed because this transit was visible from America but not visible from Europe. According to Kepler, this was to be the only transit of Venus of the seventeenth century. While the planet would come close to making a transit in 1639, by Kepler's calculations, it would just miss moving across the face of the Sun as seen from Earth: The next transit of Venus would not occur until 1761.

Kepler's calculations in this regard were the basis of Horrocks's most important discovery. Horrocks made his own precise observations of the position of Venus over a four-year period, and he observed slight deviations from the positions predicted by Kepler. In 1639, Horrocks, who had just moved to Much Hoole, near Preston, used his measurements to calculate that a transit of Venus would occur at about 3:00 P.M. on November 24, 1639, and that this transit would be visible from Much Hoole.

On the day of the predicted transit, Horrocks projected the image of the Sun onto a sheet of paper in a darkened room. Late November is not generally a time of clear skies in that part of England, but Horrocks was able to view the Sun. However, November 24, 1639, was a Sunday, which meant that, as a good Puritan, he was expected to be in church at the time of the transit. Further, since Horrocks's calculation of the exact time of the transit might be in error, as he suspected Kepler's calculation was, Horrocks intended to observe the Sun for as much of the day as possible, beginning at sunrise. He was called away several times in the morning but maintained his vigil until mid-afternoon.

The transit began at about 3:15 P.M., and Horrocks watched Venus move across the Sun for about thirty minutes before the sun set. By observing the projection of the Sun with Venus moving across its face, Horrocks was able directly to compare the angular diameter of Venus with that of the Sun. His observations showed the apparent diameter of Venus to be only 1′ 12,″ compared with the Sun's diameter of 30¢. This was much smaller than the 11′ angular diameter reported for Venus by Kepler.

In 1640, Horrocks returned to Toxteth Park, where he wrote his *Venus in sole visa* (1662; *The Transit of Venus over the Sun*, 1859) and started work on his next treatise on solar dimensions. He had begun essays on comets, tides, and the Moon as well. On January 3, 1641 Horrocks died suddenly. Much of his work was lost, but the surviving material was published much later as *Opera posthuma* (1678).

Horrocks Views Venus's Transit

Jeremiah Horrocks predicted that Venus would travel in front of the Sun at around 3:00 P.M. on November 24, 1639. He knew, however, that his calculations might be off, and he was eager to watch the Sun, projected onto a screen, for the entire day. Reproduced below is Horrocks's journal entry for that day, describing the moments leading up to his greatest discovery.

I watched carefully on the 24th from sunrise to nine o'clock, and from a little before ten until noon, and at one in the afternoon, being called away in the intervals by business of the highest importance which, for these ornamental pursuits, I could not with propriety neglect. But during all this time I saw nothing in the sun except a small and common spot. . . . This evidently had nothing to do with Venus. About fifteen minutes past three in the afternoon, when I was again at liberty to continue my labours, the clouds, as if by divine interposition, were entirely dispersed, and I was once more invited to the grateful task of repeating my observations. I then beheld a most agreeable spectacle, the object of my sanguine wishes, a spot of unusual magnitude and of a perfectly circular shape, which had already fully centred upon the sun's disc on the left, so that the limbs of the Sun and Venus precisely coincided, forming an angle of contact. Not doubting that this was really the shadow of the planet, I immediately applied myself sedulously to observe it.

Source: From "The Transit of Venus over the Sun," by Jeremiah Horrocks. Translated by Arundell Blount Whatton in his *Memoir of the Life and Labors of the Rev. Jeremiah Horrox* (London, 1859), p. 124.

Significance

Horrocks is generally regarded as the father of British astronomy. He excelled at both making precise observations and using those observations to predict planetary motion. Although Kepler recognized that transits of Mercury and Venus could be used to determine the distance from Earth to the Sun, he was unable to measure this distance because he died before the 1631 transits of Mercury and Venus took place. While Horrocks obtained only a rough estimate of the distance from Earth to the Sun, because he had observed the transit of Venus from only a single location, his value of 59 million miles (95 million kilometers) was significantly more accurate than previous estimates. It was not until a century later, following the transits of Venus in 1761 and 1769, that a more accurate value, about 93 million miles (150 million kilometers), was obtained. His results were used extensively by later astronomers, and even Sir Isaac Newton cited Horrocks's measurement of the diameter of the earth in the first edition of his *Philosophiae Naturalis Principia Mathematica* (1687; *The Mathematical Principles of Natural Philosophy*, 1729; best known as the *Principia*).

George J. Flynn

Further Reading

Aughton, Peter. *The Transit of Venus: The Brief, Brilliant Life of Jeremiah Horrocks, Father of British Astronomy*. London: George Weidenfeld & Nicholson, 2004. A two-hundred-page account of Horrocks's contributions to astronomy, focusing on his observation of the transit of Venus across the Sun.

Chapman, Allan. "Jeremiah Horrocks, the Transit of Venus, and the 'New Astronomy' in Early Seventeenth-Century England." *Quarterly Journal of the Royal Astronomical Society* 31 (1990): 333-357. A detailed account of Horrocks's research methods and his scientific accomplishments, which attempts to correct myths about Horrocks's life story.

Van Helden, Albert. *Measuring the Universe: Cosmic Dimensions from Aristarchus to Halley*. Chicago: University of Chicago Press, 1985. Includes an account of Horrocks's observations, detailing his effort to determine the distance from Earth to the Sun.

_______. *Transits of Venus: New Views of the Solar System and Galaxy*. Proceedings of IAU Colloquium 196. Cambridge, U.K.: Cambridge University Press, 2004. Includes articles by Allan Chapman and John Walton detailing Horrocks's life and his astronomical observations.

SIR FRED HOYLE

British astrophysicist

Always suspicious of received ideas and established authority, Hoyle became a flamboyant enfant terrible of science, making crucial contributions to the understanding of chemical processes within stars, advocating the steady state theory of the universe, and extending the panspermia hypothesis as an account of the origin and distribution of life in the universe.

Born: June 24, 1915; Bingley, Yorkshire, England
Died: August 20, 2001; Bournemouth, England
Also known as: Sir Frederick Hoyle (full name)
Areas of achievement: Astronomy, physics, biology

Early Life

Fred Hoyle (hoyl) was born to a wool merchant in one of the archetypal textile towns of the north of England. From Bingley Grammar School he went to Emanuel College, Cambridge, to read mathematics. He won the Mayhew Prize in 1936, the year of his graduation, and became a fellow of St. John's College in 1939. He married Barbara Clark on December 28, 1939, and the couple had two children, Geoffrey and Elisabeth.

Hoyle worked on radar systems for the British Admiralty during World War II before returning to Cambridge as a lecturer in mathematics. He had already begun work with R. A. Lyttelton on the theoretical problems of the accretion of dust and gas around large bodies. Initially recruited to apply his mathematical acumen to the models Lyttelton was building, he soon became fascinated by the problems of astrophysics and their relationship to actual astronomical data. His breakthrough as a scientist came with a 1946 paper, "The Synthesis of Elements from Carbon," from his collaboration with William A. Fowler on nuclear processes in stars. However, he did not submit a Ph.D. thesis.

Life's Work

While engaged in his war work, Hoyle met Thomas Gold and Herman Bondi, with whom he subsequently formed an axis of opposition to the theory that explained galactic red-shifts by suggesting that the universe had resulted from a primal explosion. It was Hoyle who first referred to the latter theory, disdainfully and dismissively, as "the big bang," but his scorn rebounded when the term was gladly and triumphantly adopted. Employing the same "cosmological principle of uniformity" that had led Albert Einstein to develop the theory of relativity, Hoyle, Gold, and Bondi suggested that the developing gaps between the separating galaxies must be filled in by the continuous creation of matter, thus maintaining the universe in the same stable state. Their theory was promoted in two papers published in 1948. Hoyle added mathematical substance to his collaborators' philosophical arguments.

The steady state theory maintained a viable opposition to the big bang until 1965, when the discovery of a cosmic background radiation seemed to most neutral observers to have tipped the balance conclusively in favor of the big bang. Hoyle and his collaborators initially refused to yield the point—in spite of the absence of any plausible mechanism for the spontaneous generation of matter—but gradually abandoned explicit vocal opposition. In collaboration with Indian astrophysicist Jayant Narlikar, Hoyle modified his version of the steady state model in the 1960's to incorporate local phases of rapid expansion, anticipating the later contribution of inflation theories to models of the big bang. Astronomical data provided by instruments such as the Hubble space telescope, however,

The Big Bang Gets Its Due

Ironically, the term "big bang" had been coined by its main adversary—Fred Hoyle himself—during one of his series of BBC radio talks. Hoyle used the term to belittle George Gamow's theory. Hoyle favored a different view: that the universe, although currently expanding, was infinitely old and in the long term exists in a steady state. Galaxies were not receding from each other as the aftermath of a primordial explosion (which defenders of the big bang held). Rather, space was being created between galaxies at a constant rate, and hydrogen was being created to fill that space, coalescing into nebular clouds that then formed young stars and galaxies among the old.

The problem with Hoyle's theory was that it contradicted the law of the conservation of matter: namely, that matter could neither be created nor be destroyed without being converted into energy. In the 1950's, the discovery of radio galaxies by Martin Ryle revealed that galaxies had evolved billions of years ago, supporting the big bang theory.

Once Arno A. Penzias and Robert W. Wilson discovered the cosmic microwave background radiation, Hoyle's steady state theory was largely abandoned in favor of the theory he himself had named: the big bang. Although Hoyle revised his theory to account for the background radiation, his once dominant view of the universe was out of favor. Hoyle, however, remained philosophical to the end: "The Universe eventually has its way over the prejudices of men, and I optimistically think it will do so again."

continued to provide further evidence of universal evolution, making the cosmological principle on which steady state theory had been based seem null and void. Ironically, Hoyle's work on the synthetic processes by which heavier elements are manufactured in stars made a significant indirect contribution to big bang modeling, helping to explain the manner in which the primal elements were formed in the early universe.

Hoyle was a committed popularizer of science, and he took early advantage of radio—in accordance with the cultural improvement policy adopted by the first director-general of the British Broadcasting Company (BBC), John Reith—to do work of that kind. Some of his early broadcasts were published in *The Nature of the Universe: A Series of Broadcast Lectures* (1950), which established the groundwork for such later popular texts as *A Decade of Decision* (1953), *Frontiers of Astronomy* (1955), *Man and Materialism* (1956), *Of Men and Galaxies* (1964), *Galaxies, Nuclei, and Quasars* (1965), *Man in the Universe* (1966), *The New Face of Science* (1971), and *The Cosmogony of the Solar System* (1978).

In 1957, Hoyle and Fowler coauthored another paper, this time collaborating with Geoffrey Burbidge and Margaret Burbidge. Their paper became a classic, and is best known, simply, as B^2FH. Hoyle also published his first novel, *The Black Cloud*, in that year, and became a fellow of the Royal Society. He did little formal teaching thereafter, leaving Cambridge to take up residence in the Lake District, from which he subsequently moved to the south coast. He held honorary research professorships at the University of Manchester and University College, Cardiff, however, and did a good deal of work in the United States, especially at the California Institute of Technology—where he was a visiting associate in physics in 1963—and Cornell University, where he held a visiting professorship for six years in the 1970's.

In 1958, Hoyle was appointed Plumian Professor of Astronomy and Experimental Philosophy at Cambridge. In 1967 he renewed his links with the university when he founded the Institute of Theoretical Astronomy there, and for the next six years he was a prominent member of the British scientific establishment. From 1969 to 1971 he served as vice president of the Royal Society and as president of the Royal Astronomical Society from 1971 to 1973. He also served as a member of the Science Research Council from 1967 to 1972. He was knighted in 1972.

In 1973, however, Hoyle abruptly resigned as director of the institute, severing his ties with Cambridge and other official bodies, following a dispute regarding the development of the institute's research program. He was awarded the Royal Society's Royal Medal in 1974, but was viewed thereafter with some circumspection. Fowler won the Nobel Prize for Physics in 1983, generating some controversy about whether Hoyle should have shared it, and there was much speculation as to why he had been excluded. The Royal Swedish Academy of Sciences eventually responded by giving him a prize in 1997.

While he was working at the Institute of Theoretical Astronomy, Hoyle began collaborating with Chandra Wickramasinghe, with whom he began to develop a new version of Svante August Arrhenius's panspermia hypothesis—which had suggested that life had originated elsewhere in the universe and that the earth had been "seeded" by comets. Hoyle's version proposed that cometary debris continued to rain biological material, including agents of disease, upon the planet's surface. The idea that life had originated in clouds of cosmic dust had first been exposed to public view in *The Black Cloud*, but he and Wickramasinghe developed the idea much more elaborately in such works of speculative nonfiction as *Lifecloud: The Origin of Life in the Universe* (1978), *Diseases from Space* (1979), *Evolution from Space* (1981), *The Intelligent Universe: A New View of Creation and Evolution* (1983), *Cosmic Life-Forces* (1988), and *Our Place in the Cosmos: The Unfinished Revolution* (1993). Although spectroscopic analysis of cosmic dust clouds produced increasing evidence of complex organic molecules, Hoyle's peers remained skeptical, especially of the claim that epidemics are caused by temporary deluges of extraterrestrial viruses rather than by interpersonal human infection.

Hoyle's other novels include *Ossian's Ride* (1959) and *October the First Is Too Late* (1966). With John Elliot he developed two television serials—*A for Andromeda* (1961; pb. 1962) and *The Andromeda Breakthrough* (1962; pb. 1964)—in which a radio broadcast from space provides instructions for building an advanced computer. Most of Hoyle's later fiction was written in collaboration with his son Geoffrey, and much of it was for children, extrapolating the popularizing efforts of the Molecule Club, an organization that presented educational plays and displays in schools. The more adventurous items displayed included the graphic disaster story *The Inferno* (1973) and *The Incandescent Ones* (1977). Hoyle's last solo novel, *Comet Halley* (1985), provided another vehicle for his ideas about extraterrestrial origins of life.

Significance

The determined originality of Hoyle's thinking, the mathematical ingenuity with which he extrapolated his ideas, and the stubbornness with which he defended them against criticism, are all typical of habitual rebels working against orthodoxy (and on whose efforts theoretical progress routinely depends). In two of his most sustained endeavors Hoyle appeared to back the wrong idea, but the competition he provided was a useful stimulus in each case, and the precise extent of the historical credit due him was still undecided at the beginning of the twenty-first century. It seems likely, though, that his most enduring contribution to scientific knowledge will turn out to be his analysis of heavy-element formation in stars.

Brian Stableford

Further Reading

Gough, Douglas, ed. *The Scientific Legacy of Fred Hoyle*. New York: Cambridge University Press, 2005. A collection of essays attempting to evaluate Hoyle's contributions to scientific knowledge and debate. This book lacks the coherency of Jane Gregory's work but presents various opposing viewpoints.

Gregory, Jane. *Fred Hoyle's Universe*. New York: Oxford University Press, 2005. A comprehensive summary of Hoyle's ideas relating to the origin and nature of the universe.

Lightman, Alan M., and Roberta Brawer. *Origins: The Lives and Worlds of Modern Cosmologists*. Cambridge, Mass.: Harvard University Press, 1990. An anthology that contains a long interview with Hoyle, in which his scientific ideas are critically explored.

Mitton, Simon. *Conflict in the Cosmos: Fred Hoyle's Life in Science*. Washington, D.C.: Joseph Henry Press, 2005. A biography concentrating on the controversies in which Hoyle was embroiled. This work was also published as *Fred Hoyle: A Life in Science*. London: Aurum Press 2005.

Wickramasinghe, Chandra. *A Journey with Fred Hoyle: The Search for Cosmic Life*. Hackensack, N.J.: World Scientific, 2005. A useful memoir by Hoyle's key collaborator, edited by Kamala Wickramasinghe.

EDWIN POWELL HUBBLE

American astronomer

Hubble, namesake of the Hubble Space Telescope, discovered that countless galaxies exist beyond the Milky Way, believed to have been the only galaxy of the cosmos. He also discovered the movement of those outer galaxies, disproving the centuries-old belief that the universe was static. He produced the first outline of the observable universe, revolutionizing conceptions of its size, structure, processes, and history.

Born: November 20, 1889; Marshfield, Missouri
Died: September 28, 1953; San Marino, California
Area of achievement: Astronomy

Early Life

Edwin Powell Hubble, the third of eight children born to Virginia Lee James and John Powell Hubble in Missouri, moved with his family when he was twelve years old to Wheaton, Illinois. In 1906, he won a scholarship to the University of Chicago, studying physics, chemistry, astronomy, and mathematics as well as Latin, Greek, and French. He graduated in 1910. His intellectual versatility, general curiosity, and athletic skills (he was six foot two inches) earned him considerable attention as well as a Rhodes Scholarship to Queens College, Oxford University. There, perhaps in compliance with his father's wishes, Hubble studied law, although he retained his love of astronomy. In 1913, he completed his studies at Queens, returning home to work as a legal clerk, and taught high school for a year in New Albany, Indiana.

In 1914, Hubble began work on a doctoral degree in astronomy. He sought the assistance of Forest Ray Moulton, a former astronomy professor, in obtaining a position as an assistant at the Yerkes Observatory of the University of Chicago, completing his degree in 1917. Hubble was offered a position by George Ellery Hale, director of the Mount Wilson Observatory in Southern California, to work at the observatory as an astronomer, but, instead, he applied for a commission in the Officers Reserve Corps of the U.S. Army. He attained the rank of major and served in the 343rd Infantry of the American Expeditionary Forces in France during World War I. Perhaps his military virtues, together with his Oxford education and the force of his ambition, contributed to his reputation as austere and reserved. Honorably discharged in 1919, Hubble joined the staff of the Wilson Observatory. Hale had kept the position open for him.

In 1923, Hubble's observations with the 100-inch Hooker telescope led to his discovery of Cepheid variable stars in what was then known as the Andromeda nebula. He used those stars to calculate distance and concluded that the Andromeda nebula was an independent galaxy. In 1924, *The New York Times* reported that Hubble found that spiral nebulas, which appear as whirling clouds, are stellar systems, vast in number and various in size, at great distances from the Milky Way galaxy. Hubble distinguished nebulas, luminous clouds inside the Milky Way, from extragalactic nebulas, now called galaxies. He opened a new stage in celestial inquiry, gaining international recognition. Also, in 1924, Hubble married Grace Burke, who wrote an unpublished biography of her husband, now at the Huntington Library in San Marino, California, close to Mount Wilson.

Life's Work

Hubble's natural inclinations and curiosity prompted him to decipher and analyze the riddles of the universe. In 1925, he developed a classification scheme of galaxies based on their shape or appearance. He categorized their main types as spiral, barred spiral, and elliptical, with variations of each type, and with a small percentage irregular in form. His classification system of galaxies, later known as the Hubble sequence or the Hubble Tuning Fork Diagram, remains the most widely used. He would later suggest that galactic types, forming a rough order and all existing at the same time, might indicate varying stages of change or evolution. Galaxies may collide, combine, or merge. Hubble also proposed that shapes of galaxies might result from different or decisive conditions of their formation. He could not account for the various forms of galaxies, but he developed an entire branch of astronomy devoted to their study.

Hubble pursued parallel questions of how galactic distribution and motion relate to the origin and development of the universe. By 1929, Hubble, with his assistant, Milton Humason, had evidence that the outer galaxies, other than those of the Local Group, which includes the Milky Way, were receding from each other at thousands or even hundreds of thousands of miles per second. He proposed that groups of galaxies, bound by gravity, would travel together in space. He showed a relationship between the velocities of recession of the galaxies and their distance from each other: The farther a galaxy is, the faster it appears to be receding. This distance-velocity relation is called Hubble's law.

In 1936, after a restlessly steady process of testing, modifying, and retesting, Hubble recognized more patterns in his observations and in those of other scientists and astronomers. He published his findings as *The Realm of the Nebulae* (1936), basing his views of the structure of the universe on his velocity-distance relation. Where Cepheid stars could not be detected, Hubble studied the brightness of galaxies. He employed the principle that observable lines in the spectrum of a galaxy shift toward the red end of the visible spectrum, a longer wavelength, when the lines move away; they shift toward the blue, a shorter wavelength, when an object is approaching.

Hubble's findings revealed innumerable galaxies throughout the universe, comparable to the Milky Way, and separated by unimaginable voids. He had discovered that, if considered on a large-enough scale, the universe is isotropic—that is, it has neither a center nor an edge—and that the universe is homogenous—it is generally the same in composition throughout. Furthermore, Hubble synthesized the facts and theories of other scientists in ways no one else had. He reached a staggering, awe-inspiring conclusion, involving all the bodies in space and the far reaches between them: The universe, space itself, expands at an accelerating rate in all directions. This rate of expansion is known as the Hubble constant.

Gradually, Hubble's findings, together with the work of many scientists, notably Albert Einstein, led to a new cosmology. In 1931, Einstein publicly thanked Hubble for his evidence of a dynamic universe. The evidence supported cosmological considerations in the general theory of relativity more concisely than had Einstein's own rationale. With Hubble's findings, scientists, knowing the rate at which galaxies recede, could determine the spatial distance that separates them. Furthermore, by retracing the recession, they could reconstruct the positions of the galaxies as they drew closer together: The universe, in effect, was getting smaller, denser, and younger. Astronomers inferred from the present scattering and variety of galaxies throughout space the amount of time that has passed since all the matter and energy in the universe existed in a state of utmost density. The number they reached using this technique, called the Hubble time, gave a close estimate of the age of the universe, which is believed to be 13.7 billion years. Hubble and other astronomers provided the observational reasons, and Einstein and other physicists provided theoretical reasons, for assuming an explosive origin to the universe and its subsequent expansion, popularly called the big bang.

In 1942, with the United States a part of the fighting of World War II, Hubble wanted to reassume his position in the infantry. The government, however, asked him to lead the Ballistics Research Laboratory at the Aberdeen Proving Ground in Maryland. He remained there until the end of the war, in 1945, and then returned to Mount Wilson.

At Wilson he continued his work on cosmology, another branch of astronomy he pioneered. He sought to learn whether the expansion of the universe would cease or continue forever. In 1949, he was the first to use the newly completed 200-inch Hale telescope at Mount Palomar in California.

Significance

By the time of his death from heart failure in 1953, Hubble had become the world's foremost astronomer. The Hubble crater on the moon, north of Mare Marginis, and Asteroid 2069 Hubble are named in his honor.

Hubble's picture of the universe as made up of countless and ever-changing and moving galaxies formed the basis of modern cosmology. He equipped cosmology with a new vocabulary and a new set of concepts for further study. His work initiated a breakthrough, a fundamental shift in the pattern of thought comparable to the work of Nicolas Copernicus and Galileo Galilei. The successes of the Hubble Space Telescope, first launched in 1990, testify to his legacy as well.

Timothy C. Miller

Further Reading

Christianson, Gale E. *Edwin Hubble: Mariner of the Nebulae*. New York: Farrar, Straus and Giroux, 1995. A comprehensive and detailed biography of Hubble.

Hetherington, Norriss S. *Hubble's Cosmology: A Guided* Study of Selected Texts. Tucson, Ariz.: Pachart Publishing House, 1996. An interpretative, evaluative framework for understanding some original Hubble papers. A reprint.

Hubble, Edwin. *The Realm of the Nebulae*. 1936. Reprint. New Haven, Conn.: Yale University Press, 1982. Hubble's own story of his research.

Kanipe, Jeff. *Chasing Hubble's Shadows: The Search for Galaxies at the Edge of Time*. New York: Hill and Wang, 2006. Interprets findings of the Hubble Space Telescope on the expansion of the universe.

Kron, Richard G., ed. *Evolution of the Universe of Galaxies: Edwin Hubble Centennial Symposium*. San Francisco, Calif.: Astronomical Society of the Pacific, 1990. Essays setting Hubble's work in its intellectual context.

Sandage, Allan. *The Hubble Atlas of Galaxies*. Washington, D.C.: Carnegie Institution of Washington, 1961. The final summary of Hubble's study of galaxies, prepared by a colleague.

Sharov, Alexander S., and Igor D. Novikov. *Edwin Hubble, the Discoverer of the Big Bang Universe*. Translated by Vitaly Kisin. 1993. New ed. New York: Cambridge University Press, 2005. Recounts the growth of Hubble's scientific thought and consequential discoveries.

CHRISTIAAN HUYGENS

Dutch astronomer and mathematician

Huygens was one of the greatest minds of the scientific revolution. His wave theory of light became highly influential in the nineteenth century. He discovered through his improved telescope the rings of Saturn, wrote the first formal treatise on probability, and invented the pendulum clock, which was the first truly accurate timepiece.

Born: April 14, 1629; The Hague, United Provinces (now in the Netherlands)
Died: July 8, 1695; The Hague
Areas of achievement: Astronomy, mathematics, physics, invention, science and technology

Early Life

Christiaan Huygens (KRIHS-tee-ahn HI-ghehnz) was born in The Hague. His father, Constantijn Huygens, was a diplomat in Dutch government service and a poet who is still better known in the Netherlands than his son. Constantijn attracted many notables to his house, among them the philosopher René Descartes. Christiaan and his brother were, until the age of sixteen, tutored at home in Latin, Greek, and the classics. In fact, all the learning available at that time was offered to them.

Christiaan studied for two years (1645-1647) at the nearby University of Leiden, where the supporters of the Aristotelian view of science were battling with the followers of the new Cartesian approach. Young Huygens joined the discussions before transferring to the new University of Breda, where Cartesianism was unopposed and the views of the Scholastics were not to be found. Huygens, after two years of study at Breda, had absorbed all the mathematical and scientific learning then available. He then went home and devoted himself for several years to advancing the frontiers of mathematics. He proved, for example, that the catenary formed by a chain hanging from two points was not a parabola, as Galileo had asserted.

In 1655, Huygens made his first extended visit to Paris. France and England were the countries where the scientific revolution was centered. Huygens's greatest periods of achievement were to be spent in France, where his precocious publications on mathematics and his family connections afforded him entrée into the highest intellectual circles.

Christiaan Huygens. (Library of Congress)

Life's Work

Huygens's first plunge into scientific research took place in 1655 and after, when he and his brother built improved telescopes, grinding their own lenses. With these instruments, the best made up to that time, Huygens found a satellite of Saturn, Titan, and discovered that Mars has a varied surface. He gradually discerned a ring around Saturn that nowhere touched the planet, thus improving on Galileo's more primitive observation. In order to protect the priority of this discovery while continuing his viewing, he announced by the publication of a coded message that he had found the ring.

Huygens continued his work in mathematics, which he had begun in the early 1650's with publications on hyperbolas, ellipses, and circles, and he published in 1657 the world's first formal treatise on probability.

Seeking greater scientific precision by the more accurate measurement of time, Huygens in 1656 built a pendulum clock, his greatest original invention, and with it he inaugurated modern, accurate timekeeping. Galileo had experimented with pendulums and with escapement mechanisms but had not actually constructed a clock. Huygens described the clock in his *Horologium* (1658), not to be confused with his later and greater *Horologium oscillatorium* (1673; English translation, 1966).

Huygens was preoccupied after 1660 with attempting to use his clock to solve the problem of determining longitude at sea, as the Dutch, with the world's largest merchant fleet, were intensely interested in navigational advances. Latitude could be ascertained easily by quadrant or sextant, but calculating longitude required an extremely accurate time measurement. Huygens had great hopes for his pendulum mechanism, and his marine clocks were given extensive sea trials but proved an ultimate failure in determining longitude. Not until well into the eighteenth century was the problem actually solved by the invention of a superbly accurate spring-driven chronometer. In an attempt to subject space as well as time to greater quantitative precision, Huygens built a micrometer in 1658. With it he could establish, within a few seconds of arc, the position of a heavenly body.

In 1661, the year he joined the Royal Society, Huygens returned to Paris, where King Louis XIV's chief minister, Jean-Baptiste Colbert, eager to retain Huygens's services for the French, procured for the Dutchman a significant government grant for scientific work.

Scholars interested in science had been meeting in Paris for years in the salons, or drawing rooms, of intellectuals and the wealthy. The Crown wished to formalize such gatherings, and so it founded the Académie Royale des Sciences (Royal Academy of Sciences) in 1666. Huygens's scientific reputation was so formidable that he was made a charter member of the academy, given a regular salary larger than that of any other member, and handed the keys to an apartment in the Bibliothèque Royale. Thus, he commenced, in 1666, a period of residence in Paris lasting until 1681. Except for two extended visits home to The Hague, Huygens remained in Paris. Still, when Huygens left Paris in 1681 for a third trip to the Netherlands, he never returned. His patron Colbert died in 1683, and anti-Protestant sentiment was growing in France, making Huygens's position difficult, as he was nominally a Calvinist.

Huygens's philosophy of science was intermediate between those of the two giants of his day: René Descartes in France and Sir Isaac Newton in England. Huygens grew up a Cartesian but broke with his mentor over the latter's extreme devotion to the mathematical, or deductive, approach to science. Descartes attempted to explain all phenomena by use of deductive logic alone. Newton, on the contrary, relied on observations and experiments as the bases for his laws.

Huygens's basic approach to the universe was mechanistic, an impact or billiard ball physics, in which he denied all action at a distance. He did prefer, however, Descartes's supposedly more tangible "vortices" of "subtle matter" to Newton's "gravity" in explaining the movements of heavenly bodies. Gravity worked unseen and over distance, moving bodies without apparently touching them. In the

Huygens Considers Life on Other Planets

Christiaan Huygens was one of the first scientists to consider the possibility of life on other planets. In his work Cosmotheoros *(1698), published the same year in English, he writes about the inhabitants ("the Planetarians"), social composition, flora and fauna, weather, and built landscapes of these other worlds.*

A Man that is of *Copernicus*'s Opinion, that this Earth of ours is a Planet, carried round and enlightened by the Sun, like the rest of them, cannot but sometimes have a fancy, that it's not improbable that the rest of the Planets have their Dress and Furniture, nay and their Inhabitants too as well as this Earth of ours.

There may arise another Question, whether there be in the Planets but one or more sorts of rational Creatures possessed of different degrees of Reason and Sense.

But some body may perhaps object, and that not without reason at first sight, that the Planetarians it's likely are destitute of all refined Knowledge, just as the Americans were before they had Commerce with the Europeans.

What I am now going to say may seem somewhat more bold. . . . For if these new Nations live in Society, as I have pretty well showed they do, 'tis somewhat more than probable that they enjoy not only the Profit, but the Pleasures arising from such a Society: such as Conversation, Amours, Jesting, and Sights.

[W]e have good reason to believe they build themselves Houses, because we are sure they be not without their Showers. . . . To protect themselves from these [weather conditions], and that they may pass their Nights in quiet and safety, they must build themselves Tents or Huts, or live in holes of the earth.

Source: Huygens, *The Celestial Worlds Discovered* (London: Cass, 1968), pp. 1-2, 56, 66, 80, 81. Rendered into modern English by Desiree Dreeuws.

matter of relativity, however, Huygens was in advance of Newton. For Huygens, all motion in the universe was relative. Huygens, in this regard, was closer to the later work of Albert Einstein.

Huygens and Newton also differed over what constituted the nature of light. Newton considered light to be composed of particles or corpuscles emitted in steady streams from a light source; Huygens regarded the transmission of such particles through empty space to be mere Newtonian "action at a distance" again and incompatible with a mechanistic view of nature. Huygens propounded a wave theory of light, maintaining that a medium, an ether, must exist in space and that light is transmitted with a rapid but finite speed as shock waves in this medium. The ether, he believed, was composed of tiny, closely spaced, elastic particles, which vibrate and pass on the waves of light. Thus he did not view light itself as a substance as did Newton, nor did he consider it to be instantaneously transmitted as did Descartes.

Huygens remained in communication with Newton, although his relations with the Royal Society dwindled after 1678. He visited England again in 1689, conversing with Newton and addressing the Royal Society on his non-Newtonian theory of gravity, a theory published the following year as *Discours de la cause de la pesanteur* (1690; discourse on the cause of gravity). Huygens's last years were spent in The Hague, where he died in 1695.

Significance

Huygens would have been a great intellect in any age, but the magnitude of his brilliance was not as apparent as it might have been had he not been so close in space and time to such luminaries as Newton and Descartes. Nevertheless, his achievements were considerable.

As an astronomer he not only discovered the rings and a satellite of Saturn but also was the first to notice the nebula in the constellation Orion. In his work on centrifugal force, he was the first to theorize that Earth must be an oblate spheroid. He worked with microscopes as well as telescopes and translated some of the letters of the great Dutch microscopist Antoni van Leeuwenhoek. His work in mechanics of systems led him to invent the pendulum clock, the world's first truly accurate timepiece.

Huygens originated the wave theory of light and thereby established the science of physical optics, although the wave theory was not accepted in his own century or the next one as the fundamental explanation of the nature of light. In the seventeenth and eighteenth centuries, Newton's theory of the corpuscular or particulate nature of light held sway, but nineteenth century scientists focused on the diffraction of light, which can be explained only by wavelength. Thus, the ether enjoyed a renewed popularity, and in the nineteenth century Huygens's theory was the prevailing view.

In the early twentieth century, the physics of Max Planck and Einstein led to a synthesizing of Newton's and Huygens's views into the quantum theory, in which both concepts were correct. Light today is considered indeed to have various wavelength properties, but it is also seen as moving in packets of energy called photons. A mathematical genius, Huygens improved on the work of Galileo in mechanics and astronomy, and he conferred and contested with the giants of his own generation. Always the pure scientist, he never dabbled in the metaphysical and was uninfluenced by religion—either the Calvinism of the Netherlands or the Catholicism of France. He died as he had lived, sure of only one thing: that there is in this universe no ultimate certainty.

Allan D. Charles

Further Reading

Bell, Arthur E. *Christian Huygens and the Development of Science in the Seventeenth Century*. New York: Longmans, Green, 1947. This scholarly, well-written volume has long been the standard biography of Huygens in the English language. Bell gives a thorough and interesting account of Huygens's life, his theoretical approach to science, and his actual scientific work.

Chappell, Vere, ed. *Seventeenth-Century Natural Scientists*. New York: Garland, 1992. A collection of essays about Huygens, Robert Boyle, and Sir Isaac Newton, describing their individual scientific contributions and the wider philosophical and intellectual world in which they worked. Also explores the influence of Descartes on their conceptions of scientific method.

Dijksterhuis, Fokko Jan. *Lenses and Waves: Christiaan Huygens and the Mathematical Science of Optics in the Seventeenth Century*. Dordrecht, the Netherlands: Kluwer Academic, 2004. Intended for a reader with some knowledge of physics, this work provides an overview of Huygens's contributions to the science of optics and a broader discussion of that science during the seventeenth century.

Elzinga, Aant. *On a Research Program in Early Modern Physics*. New York: Humanities Press, 1972. Originally a doctoral dissertation, this work contains a lengthy chapter devoted to Huygens's theory of research and how he broke with the system of his mentor, Descartes.

Huygens, Christiaan. *The Celestial Worlds Discovered: Or, Conjectures Concerning the Inhabitants, Plants, and Productions of the Worlds in the Planets*. London: Childe, 1698. Reprint. London: Cass, 1968. Short, nontechnical, and readable, this continually popular book gives excellent insight into Huygens's thinking. He was one of the first scientists to conjecture about life on other planets.

Simmons, George Findlay. *Calculus Gems: Brief Lives and Memorable Mathematics*. New York: McGraw-Hill, 1992. Huygens is one of the mathematicians included in this collection of brief biographies, which discusses his mathematical theories and his relationship with Gottfried Wilhelm Leibniz, his mathematics student.

Struik, Dirk J. *The Land of Stevin and Huygens: A Sketch of Science and Technology in the Dutch Republic During the Golden Century*. Boston: D. Reidel, 1981. A short, illustrated volume, this book centers on Huygens as the chief claim to fame of the Netherlands in the scientific revolution of the seventeenth century. Struik not only describes Huygens's scientific work but also shows how Huygens was influenced by technological demands and commercial pressures.

Tabak, John. *Probability and Statistics: The Science of Uncertainty*. New York: Facts On File, 2004. The chapter about the nature of chance includes a brief discussion of Huygens's ideas on this subject.

Yoder, Joella G. *Unrolling Time: Huygens and the Mathematization of Nature*. New York: Cambridge University Press, 1988. This book describes how Huygens used mathematics to substantiate his mechanistic view of the universe. Details Huygens's discoveries, including his invention of the pendulum clock, and describes how these discoveries are linked to his mathematical universe.

JOHANNES KEPLER

German astronomer

Through the application of his exceptional intellect, faith, and tenacity, Kepler created the science of modern astronomy—the fusing together of physics and astronomy—and provided the solid foundation upon which Isaac Newton built his laws of universal gravitation. He also is considered the father of modern optics.

Born: December 27, 1571; Weil, Swabia (now Weil der Stadt, Germany)
Died: November 15, 1630; Regensburg, Bavaria (now in Germany)
Areas of achievement: Astronomy, physics, mathematics, science and technology

Early Life

Johannes Kepler (yoh-HAHN-ehs KEHP-lehr) was born prematurely, the first child of Heinrich and Katharine Kepler, and his childhood was exceptionally difficult. He was small and sickly, with thin limbs and a large, pasty face surrounded by dark hair. It is one of the ironies of history that the child who would one day revolutionize astronomy had poor eyesight. His mother was a small woman with a nasty disposition; his father appears to have had no established trade and, in 1574, simply left the family to fight for the Catholic duke of Alva in the Netherlands. His mother followed a year later, leaving Kepler in the care of his grandparents, who treated him badly. In 1576, his parents returned, but this provided only a dubious improvement in his family life.

Under such circumstances, it is not surprising that Kepler's self-image as a child was terrible; he described himself once as a "mangy dog." Possibly in response to the instability in his life, Kepler developed a pronounced religious disposition at a young age. Indeed, of all his childhood memories only two stood out as pleasant. When he was six, Kepler's mother took him to a hill to see a comet, and at age nine his parents took him outside to observe an eclipse of the Moon. The seeds were planted that would influence the direction of his life.

Yet it was obvious even to his parents that Kepler was a bright child, and school provided a way for him to divert himself from the suffering of family life while building his self-assurance and developing his intellect. Kepler was fortunate that the Lutheran dukes of Württemberg provided generous educational scholarships for the intelligent sons of poor parents.

Kepler began his schooling at age seven at the Latin school in Leonberg. He completed the curriculum at age twelve, and his intellect, poor health, and pious nature preordained him to a clerical career. After passing a competitive exam on October 16, 1584, Kepler continued his studies at the higher seminary at Maulbronn. On September 17, 1589, he entered the University of Tübingen, where he was particularly influenced by his professor of astronomy, Michael Mästlin. Mästlin was unique in that he believed that Nicolaus Copernicus's

Johannes Kepler. (Library of Congress)

heliocentric theory was essentially true. Influenced by this exceptional teacher, Kepler accepted the Copernican view of the universe—an act that would have a profound impact on his life and the future of scientific thought.

At the age of twenty, Kepler was matriculated at the Tübingen Theological School. His career path seemed assured, but circumstances fatefully intervened. In 1593, the mathematician of the Lutheran high school of Graz died, and the school requested Tübingen to recommend a replacement. Kepler was nominated and agreed to accept the position.

His job was to teach mathematics and to publish the annual calendar of astrological forecasts. Kepler was lucky with his first calendar, correctly predicting a cold spell and a Turkish invasion. He would be involved with astrology all of his professional life, but he did it primarily to supplement his income. Still, while he considered popular astrological forecasts a "dreadful" superstition, Kepler believed that astrology could become an exact empirical science, and with that aim in mind, he would write several treatises on the subject.

Kepler was a terrible teacher. He was often unintelligible, launching into obscure digressions whenever a thought occurred to him. Yet it was during one of these lectures that the event happened that set Kepler on the path that would ultimately lead him to reinterpret the prevailing view of the universe.

Life's Work

Kepler found his first year at Graz very trying. To escape its frustrations, he turned to the astronomical studies he had experimented with at Tübingen. The more he contemplated the Copernican system, the more he became convinced of its truth, but he was also aware that it was not the final, definitive explanation of the operation of the universe.

Already in 1595, Kepler was beginning to ask the questions that would determine that course of his scientific inquiry. Why are there only six planets? What determines their distances from the sun? Why do the planets move more slowly the farther away they are from the sun? These and other queries had been coursing through his mind when, on July 19, 1595, as he was teaching, an incredible thought struck him. On the blackboard, he had drawn a figure showing an outer circle circumscribing a triangle, which enclosed an inner circle. As Kepler looked at the two circles, he was suddenly dumbfounded by the realization that the ratios of the two circles were the same as those of the orbits of Saturn and Jupiter. Could it be that there were only six planets because planetary orbits were related to the five regular solids—the tetrahedron, cube, octahedron, dodecahedron, and icosahedron—of Euclid's geometry?

This revelation was the basis of Kepler's first major work, *Mysterium cosmographicum* (1596; *The Secret of the Universe*, 1981). His thesis was that one of the five regular solids fit between each of the invisible spheres that carried the six planets. The key insight of this book, however, revolved around his search for a mathematical relationship between a planet's distance from the Sun and the time necessary for it to complete its orbit. Kepler concluded that there must be a force emanating from the Sun that swept the planets around their orbits. The outer planets moved more slowly because this force diminished in a ratio to distance, just as light did. Here is found the first hint of celestial mechanics, the joining of physics and astronomy, that would lead to the laws of planetary motion.

Kepler sent copies of the work to a number of scientists in Europe, including Tycho Brahe, who had spent several years making painstakingly accurate observations of planetary orbits and who would shortly become the imperial mathematician of the Holy Roman Empire. Although he did not agree with the Copernican underpinning of Kepler's work, Brahe was impressed by Kepler's knowledge of mathematics and astronomy, and he invited Kepler to join his staff in the observatory at Benatek, just outside Prague. Kepler was flattered by the invitation, but he had just been married and was too poor to afford the trip. The deteriorating religious situation in Graz, in which Protestants were being forced either to convert to Catholicism or to emigrate, compelled Kepler to make a decision. In 1600, Kepler joined Brahe in Prague, where he would remain until 1612. This period would be the most productive of his life.

Kepler's first publication after arriving at Prague, *De fundamentis astrologiae certioribus* (1601; the more reliable bases of astrology), while rejecting the belief that celestial bodies direct people's lives, supported the mystical view that there is a harmony between the universe and the individual. In 1606, after observing a supernova, Kepler published *De stella nova in pede serpentarii* (the new star in the foot of the serpent bearer), which argued that the universe of fixed stars was not pure and changeless, as had been believed. In 1604, Kepler authored a major work on optics, *Ad Vitellionem paralipomena, quibus astronomiae pars optica traditur* (*Optics: Paralipomena to Witelo and Optical Part of Astronomy*, 2000). The main subject was the atmospheric refraction, or bending, of light as it enters Earth's atmosphere from space. An understanding of this phenomenon was vital if Kepler was to make optimum use of Brahe's observational data. In this work, Kepler for the first time explained the fundamental structure and function of the human eye. Although he was unsuccessful in developing the law of refraction, he was able to create an improved table of refraction.

Kepler concluded his work in optics with *Dioptrice* (1611; partial translation of the preface, 1880), which not only restated his concept of refraction but also covered such subjects as reflections, images, magnification, and the optical principles of the astronomical telescope. These two works have justifiably established Kepler as the father of modern optics. That same year, Kepler wrote the *Dissertatiocum nuncio sidereo* (1610; *Kepler's Conversation with Galileo's Sidereal Messenger* 1965), placing his then-considerable scientific prestige in support of Galileo's astronomical discoveries. Kepler's great magnum opus, published in 1609 after years of painstaking effort, was the *Astronomia nova* (*New Astronomy*, 1992).

When Kepler first arrived at Prague in 1600, Brahe immediately assigned him to investigate the orbit of Mars, which Brahe and his assistant, Danish astronomer Longomontanus (Christian Severin), had been unable to determine. The selection of Mars was particularly fortuitous because of all the planets, Mars has the most elliptical orbit and, therefore, provided the best opportunity for discovering the secrets of planetary motion. Once Kepler had full use of Brahe's observational data following Brahe's death in 1601—for now Kepler was the imperial mathematician—he became completely immersed in this project.

As he began the study of Mars, Kepler had a thoroughly Copernican concept of planetary motion. Each planet revolved around the Sun at a uniform speed in a perfectly circular orbit, considered the perfect geometrical shape. Accordingly, Kepler first tried to show that the Martian orbit was circular. After three years of intense, repetitive calculations, Kepler believed that he had proved Mars had a circular orbit, only to discover that two of Brahe's innumerable observations of the planet differed from Kepler's orbit by only 8 minutes of arc. (The width of a pinhead held at arm's length approximately equals 8 minutes of arc.) Previous scientists would have made the evidence fit the theory, but Kepler would not. Believing that God would not create anything imperfect, Kepler knew that he could not ignore those 8 minutes of arc. The orbit of Mars could not be circular but had to be some other geometrical curve.

Kepler now had to redetermine the Martian orbit, but first he had to recalculate the orbit of Earth. Earth was his observatory, and if there were any misconceptions regarding its motion, then all conclusions regarding other planetary motion would be in error. He discovered that Earth did not revolve around the Sun at a uniform speed, but rather moved faster or slower depending upon its distance from the sun. Obviously, the Platonic and Copernican concept of uniform motion was incorrect. Kepler, however, discovered a new type of uniform motion: With the Sun as its focus, the planet, while revolving along the periphery of its orbit, will sweep out, in equal intervals of time, equal areas of the orbit, and unequal arcs along the periphery of the orbit.

The second law determined the variations of the planet's speed along its orbit but not the shape of the orbit itself. For two more years, Kepler worked on that problem before coming upon the solution: an ellipse. This led to his first law (discovered after his second law): A planet moves in an ellipse with the Sun at one of the two foci.

Kepler's third law was not published until 1619 in the *Harmonices mundi* (partial translation as *Harmonies of the World*, 1952). Here Kepler states that the squares of the planetary periods (the time it takes a planet to complete its orbit) are proportional to the cubes of their average distance from the Sun (the 3/2 ratio). The farther from the Sun an object is, the slower it moves.

These two works, the *New Astronomy* and *Harmonies of the World*, completed Kepler's work on planetary motion and provided the basis for Isaac Newton's explanation of universal gravitation, which affects every material object in the universe.

In the meantime, Kepler's life underwent much upheaval. In 1610, his wife died, and the next year his patron, Holy Roman Emperor Rudolf II, slipped into insanity and was deposed by his brother, Matthias, who became the new king of Hungary. Although he was reappointed imperial mathematician, Kepler left Prague and moved to Linz, Austria. In 1623, he was married to Susanna Reuttinger, who although several years younger than Kepler and of lower social status, proved to be one of the few joys of his later life.

While in Linz, Kepler published his *Epitome astronomiae Copernicanae* (1618-1621; partial translation as *Epitome of Copernican Astronomy*, 1939). This relatively unknown and underrated work is highly significant, ranking next to Ptolemy's *Almagest* (second century) and Copernicus's *De revolutionibus* (1543) as the first systematic elaboration of the concept of celestial mechanics established by Kepler. In it, he not only conclusively proved the validity of the Copernican view of the universe but also revealed all the knowledge he had uncovered during his many years of research.

As he worked on *Epitome of Copernican Astronomy*, Kepler heard that his mother was going to be tried as a witch, a capital offense. Self-interest as well as familial devotion led him to conduct a successful defense. Had she been convicted, not only would there be severe consequences for her, including possible death, but Kepler's status as imperial mathematician also could have been imperiled.

Kepler also planned to publish *Tabulae Rudolphinae* (English translation, 1675), named in honor of Rudolf II, in Linz, but a peasants' rebellion and religious unrest forced Kepler to move to Ulm, where the work was first published in 1627. Based on Brahe's observations, *Tabulae Rudolphinae* is a book on practical astronomy. It became an indispensable astronomical tool for more than a century, and Kepler considered it the crowning achievement of his life.

The last three years of Kepler's life were a struggle. Albrecht Wenzel von Wallenstein, the famous mercenary general of the Thirty Years' War (1618-1648), promised to meet Kepler's financial needs, but the general was unreliable. In 1628, Kepler moved to Zagań, Silesia, and late in 1630 he left for Austria to collect some funds he was owed. He stopped at Regensburg, where the Imperial Diet was meeting, fell ill, and died on November 15, 1630. Scientists throughout Europe mourned his death.

Significance

Kepler's impact on the development of astronomy and general science was enormous. By the sheer force of his intellect and the tenacity of his spirit, he forged ahead in the understanding of the cosmos, further than any of his contemporaries. Kepler not only provided the mathematical proof of the Copernican system but also went far beyond it, creating the science of modern astronomy, in which physics, the concept of physical force, and astronomy were fused together. He discovered the famous three laws, created the science of optics, and came very close to discovering gravity. His determination that the theory must fit the facts and not vice versa established the standard for future scientific inquiry. Without Kepler, there would not have been Newton's laws of universal gravitation.

Kepler also had a more abstract impact on Western society. A society's perception of the cosmos is reflected in the way it views itself. By demonstrating conclusively for the first time that the universe operates according to fixed, natural laws, Kepler assisted Western society in freeing itself from the shackles of superstition and ignorance.

Basic to Kepler's success was his Christian faith. Although his belief in the harmony of the worlds led him into many mystical conjectures, Kepler's firm belief that God would only have created a harmonious universe, where there had to be a predetermined reason for the occurrence of certain events, provided the proper attitude for discerning the existence of natural laws. Indeed, Kepler's faith, rather than being a hindrance, was a creative force pushing him ever forward.

Ronald F. Smith

Further Reading

Armitage, Agnus. *John Kepler*. London: Faber & Faber, 1966. Although not as detailed as other biographies, this is a very good introduction to Kepler and his work. It is lucid, includes excellent illustrations, and provides clear explanations of Kepler's calculations.

Bishop, Philip W., and George Schwartz, eds. *Moments of Discovery*. Vol. 1 in *The Origins of Science*. New York: Basic Books, 1958. Pages 265-277 contain the first partial translation of *Astronomia nova*, specifically Kepler's explanation of his first law.

Caspar, Max. *Kepler*. New York: Abelard-Schuman, 1959. This translation of the 1947 German edition is the definitive work on Kepler, thoroughly investigating every aspect of his life and work. Well written and lucid, this work is a must for any study of Kepler.

Connor, James A. *Kepler's Witch: An Astronomer's Discovery of Cosmic Order Amid Religious War, Political Intrigue, and the Heresy Trial of His Mother*. San Francisco, Calif.: HarperSanFrancisco, 2004. A biography of Kepler, placing him within the context of seventeenth century life. Depicts him as one deeply driven by his Lutheran faith and one who was trying to survive the politics of the Counter-Reformation.

Gilder, Joshua, and Anne-Lee Gilder. *Heavenly Intrigue: Johannes Kepler, Tycho Brahe, and the Murder Behind One of History's Greatest Scientific Discoveries*. New York: Doubleday, 2004. Depicts the troubled relationship of Kepler and his mentor, Brahe, contending Kepler murdered Brahe to obtain the scientific data Kepler needed to complete his laws of planetary motion. Portrays Kepler as a virtual sociopath, consumed with anger and beset with illness.

Kepler, Johannes. *New Astronomy*. Translated by William H. Donahue. New York: Cambridge University Press, 1992. The first full translation in English of Kepler's major work. Includes a bibliography and an index.

_______. *Optics: Paralipomena to Witelo and Optical Part of Astronomy*. Translated by William H. Donahue. Santa Fe, N.Mex.: Green Lion Press, 2000. The first English translation of Kepler's major work on optics. Includes a bibliography and an index.

Koestler, Arthur. *The Watershed: A Biography of Johannes Kepler*. Garden City, N.Y.: Anchor Books, 1960. This Science Study series volume is a fine introduction to Kepler and his work. The discussion on Kepler and Galileo is excellent, showing that Kepler had a much greater impact on the development of astronomy and physics than Galileo.

Small, Robert. *An Account of the Astronomical Discoveries of Kepler*. London: J. Mawman, 1804. Reprint. Madison: University of Wisconsin Press, 1963. Until the translation of Caspar, this was the best work on Kepler in English. It remains the best English work for readers interested in the mathematical details of Kepler's theories. Contains reproductions of many of Kepler's geometric figures.

Spiller, Elizabeth. *Science, Reading, and Renaissance Literature: The Art of Making Knowledge*. New York: Cambridge University Press, 2004. Studies literary works and the works of Kepler and other scientists to examine the changing disciplines of literature and science during the Renaissance.

PIERRE-SIMON LAPLACE

French mathematician

Laplace made groundbreaking mathematical contributions to probability theory and statistical analysis. Using Isaac Newton's theory of gravitation, he also performed detailed mathematical analyses of the shape of the earth and the orbits of comets, planets, and their moons.

Born: March 23, 1749; Beaumont-en-Auge, Normandy, France
Died: March 5, 1827; Paris, France
Also known as: Marquis de Laplace
Areas of achievement: Astronomy, mathematics, physics

Early Life

Pierre-Simon Laplace (lah plahs) was born into a well-established and prosperous family of farmers and merchants in southern Normandy. An ecclesiastical career in the Church was originally planned for Laplace by his father, and he attended the Benedictine secondary school in Beaumont-en-Auge between the ages of seven and sixteen. His interest in mathematics blossomed during two years at the University of Caen, beginning in 1766.

In 1768, Laplace went to Paris to pursue a career in mathematics; he remained a permanent resident of Paris or its immediate vicinity for the rest of his life. Soon after his arrival in Paris, he sought and won the patronage of Jean Le Rond d'Alembert, a mathematician, physicist, and philosopher with great influence among French intellectuals. D'Alembert found Laplace employment teaching mathematics to military cadets at the École Militaire, and it was in this position that Laplace wrote his first memoirs in mathematics and astronomy.

In 1773, Laplace was elected to the Academy of Sciences as a mathematician. This achievement, at the relatively young age of twenty-four, was based upon the merits of thirteen memoirs he had presented to academy committees for review. Some of Laplace's earliest mathematical interests involved the calculation of odds in games of chance. At a time when there was not yet a field of mathematics devoted to the systematic study of probability, Laplace played a major role in carrying the early development of this topic beyond the rules of thumb of gambling and the preliminary conclusions of earlier mathematicians. In addition, Laplace emphasized the relevance of probability to the analysis of statistics. He believed that, because all experimental data are imprecise to some extent, it is important to be able to calculate an appropriate average or mean value from a collection of observations. Furthermore, this mean value should be calculated in such a way as to minimize its difference from the actual value of the quantity being measured.

Statistical problems of this type inspired Laplace's initial interest in astronomy. He became intrigued by the process through which new astronomical data should be incorporated into calculations of probabilities for future observations. In particular, he concentrated on the application of Sir Isaac Newton's law of gravitation to the motions of the comets and planets. Laplace's interest in physics thus had a strong mathematical orientation. Throughout his career, he retained his early concentration on the solution of problems suggested by the mathematical implications of physical laws; he never devoted himself to extensive experimental investigation of new phenomena. Laplace's primary motivation was a deep conviction that, even if human limitations prevent an exact knowledge of natural laws and experimental conditions, it is still possible progressively to eliminate error through increasingly accurate approximations.

Very little is known about Laplace's personal life during these early years. He does not seem to have stimulated strong friendship or animosity. In 1788, he married Marie-Charlotte de Courty de Romanges, who was twenty years younger than himself, and they had two children. Laplace established and maintained comfortable but disciplined living habits, and he retained an undiminished mental clarity to the moment of his death.

Life's Work

Although a brief summary of Laplace's life's work requires some classification by topics and an emphasis on final results rather than chronology, the highly integrated and developmental nature of his research should not be forgotten. For example, mathematical techniques that he invented for the solution of problems in probability theory often were immediately applied to similar problems in physics or astronomy. Because Laplace was particularly interested in approximate or probable solutions and the analysis of error, he repeatedly revised his mathematical techniques to accommodate new data.

Laplace's contributions to probability theory were both technical and philosophical. This twofold concern is expressed in the titles of the influential volumes in which he summarized his work, *Théorie analytique des probabilités* (1812; analytic theory of probability) and *Essai philosophique sur les probabilités* (1814; *A Philosophical Essay on Probabilities*, 1902).

The *Théorie analytique des probabilités* was the first comprehensive treatise devoted entirely to the subject of probability. Laplace provided a groundbreaking, although necessarily imperfect, characterization of the techniques, subject matter, and practical applications of the new field. He relied on the traditional problems generated by games of chance, such as lotteries, to motivate his mathematical innovations, but he pointed toward the future by generalizing these methods and applying them to many other topics. For example, because the calculation of odds in games of chance so often requires the summation of long series of fractions in which each term in the series differs from the others according to a regular pattern, Laplace began by reviewing some of the methods he had discovered to approximate the sums of such series, particularly when very large numbers are involved.

Laplace then proceeded to state what has since come to be called Bayes's theorem, after an early predecessor of Laplace. This theorem states how to use partial or incomplete information to calculate the conditional probability of an event in terms of its absolute or unconditional probability and the conditional probability of its cause. Laplace was one of the first to make extensive use of this theorem; it was particularly important to him because of its relevance to how calculations of probability should change in response to new knowledge.

The *Théorie analytique des probabilités* includes Laplace's applications of his mathematical techniques to problems generated by the analysis of data from such diverse topics as census figures, insurance rates, instrumentation error, astronomy, geodesy, election prognostication, and jury selection. In particular, he gave an important statement of what has since been called the least square law for the calculation of a mean value for a set of data in such a way that the resulting error from the true value is minimized.

A Philosophical Essay on Probabilities has been one of Laplace's most widely read works; it includes the conceptual basis upon which Laplace constructed his mathematical techniques. Most important, Laplace stated and relied upon a definition of probability that has been a source of considerable philosophical debate. Given a situation in which specific equally possible cases are the results of various processes (such as rolling dice) and correspond to favorable or unfavorable events, Laplace defined the probability of an event as the fraction formed by dividing the number of cases that correspond to or cause that event by the total number of possible cases. When the cases in question are not equally possible (as when dice are loaded), the calculation must be altered in an attempt to include this information. Laplace's definition thus calls attention to his treatment of probability as an application of mathematics made necessary only by human ignorance.

In one of the most famous passages in *A Philosophical Essay on Probabilities*, Laplace expresses this view by describing a supreme intelligence with a complete knowledge of the universe and its laws at any specific moment; for such an intelligence, Laplace believed that probability calculations would be unnecessary because the future and past could be calculated simply through an application of the laws of nature to the given perfectly stipulated set of conditions. Because knowledge of natural laws and the state of the world is always limited, probability is an essential feature of all human affairs. Nevertheless, Laplace's emphasis was not on the negative aspect of this conclusion but on the mathematical regularities to which even seemingly arbitrary sequences of events conform.

The domain in which Laplace saw the closest human approach to the knowledge of his hypothetical supreme intelligence was the application of Newton's theory of gravitation to the solar system. Since Newton's publication of his theory in 1686, mathematicians and physicists had reformulated his results using increasingly sophisticated mathematics. By Laplace's time, Newton's theory could be stated in a type of mathematics known as partial differential equations. Laplace made major contributions to the solution of equations of this type, including the famous technique of "Laplace transforms" and the use of a "potential" function to characterize a field of force.

Laplace made remarkably detailed applications of Newton's results to the orbits of the planets, moons, and comets. Some of his most famous calculations involve his demonstration of the very long-term periodic variations in the orbits of Jupiter and Saturn. Laplace thus contributed to an increasing knowledge of the stability and internal motions of the solar system. He also applied gravitation theory to the tides, the shape of Earth, and the rings of Saturn. His hypothesis that the solar system was formed through the condensation of a diffuse solar atmosphere became a starting point for more detailed subsequent theories.

Pierre-Simon Laplace. (Library of Congress)

Newtonian gravitation theory became Laplace's model for precision and clarity in all other branches of physics. He encouraged his colleagues to attempt similar analyses in optics, heat, electricity, and magnetism. His influence was particularly strong among French physicists between 1805 and 1815. By his death in 1827, however, this attempt to base all physics upon short-range forces had achieved only limited success; aside from the mathematical methods he developed, Laplace's conceptual contributions to physics were not as long-lasting as his more fundamental insights in probability theory.

Significance

Pierre-Simon Laplace's cultural influence extended far beyond the relatively small circle of mathematicians who could appreciate the brilliant technical detail in his work. In several ways he has become a symbol of some important aspects of the rapid scientific progress that took place during his career as a result of his role in institutional changes in the scientific profession and the implications that have been drawn from his conclusions and methods.

Laplace was very active within the highly centralized French scientific community. As a member of the French Academy of Sciences, he served on numerous research or evaluative committees that were commissioned by the French government. For example, following the French Revolution in 1789, he was an influential designer and advocate of the metric system, which has become the most widely used international system of scientific units. The academy was disbanded during the radical phase of the Revolution in 1793, but in 1796 Laplace became the president of the scientific class of the new Institute of France.

Highly publicized institute prizes were regularly offered for essays in physics and mathematics, and Laplace exerted a powerful influence on French physics through the attention he devoted to choice of topic and support for his preferred candidates. He also played an important part in the early organization of the École Polytechnique, the prestigious school of engineering founded in 1795. Although Laplace lived through turbulent political changes, he remained in positions of high scientific status through the Napoleonic era and into the Bourbon Restoration, when he was raised to the nobility as a marquis. Laplace seems to have held few strong political views, and he thus is sometimes cited as an example of a powerful scientist indifferent to social or political conditions.

Aside from his work in probability and statistics, which has quite direct impact on modern societies, other aspects of Laplace's work have contributed to general perceptions of the goals, limitations, and methods of science. With Newton's theory of gravitation as his model, Laplace was convinced that, although human knowledge of nature is always limited, there are inevitable regularities that can be expressed approximately with ever-increasing accuracy. Laplace thus has become a symbol of nineteenth century scientific determinism, the view that the uncertainty of the future is only the result of human ignorance of the natural laws that determine it in every detail.

When Napoleon I asked Laplace why God did not play a role in Laplace's analysis of the stability of the solar system, Laplace replied that he had had no need for such a hypothesis. Laplace thus contributed to a growing association of the scientific tradition with atheism and materialism. Finally, Laplace's style of mathematical physics has become a primary example of a reductionistic research strategy. Just as the gravitational effect of a large mass is determined by the sum of the forces exerted by all of its parts, Laplace expected all phenomena to reduce to collections of individual interactions. His success in implementing this method contributed to widespread perceptions that this is a necessary component of scientific investigation.

James R. Hofmann

Further Reading

Arago, François. "Laplace." In *Biographies of Distinguished Scientific Men.* New York: Ticknor & Fields, 1859. Arago was a student and colleague of Laplace for many years. His essay discusses only Laplace's work in astronomy and concentrates on his study of the stability of the solar system.

Brush, Stephen G. *The Origin of the Solar System and the Core of the Earth from Laplace to Jeffreys: Nebulous Earth.* Vol. 1 in *A History of Modern Planetary Physics.* New York: Cambridge University Press, 1996. Traces the evolution of Laplace's nebular hypotheses, the most popular nineteenth century explanation for the origin of the solar system.

Fox, Robert. "The Rise and Fall of Laplacian Physics." *Historical Studies in the Physical Sciences* 4 (1974): 89-136. This is an excellent summary of Laplace's efforts to direct French physics according to a research program based upon short-range forces.

Gillespie, Charles Coulston, Robert Fox, and Ivor Grattan-Guiness. *Pierre-Simon Laplace, 1749-1827: A Life in Exact Science.* Princeton, N.J.: Princeton University

Press, 1997. Focuses on Laplace's research program and his work with the Academy of Science. Includes biographical information from a scientific point of view, a description of Laplace's efforts to gather young physicists who would work with the Newtonian model in physics, and an overview of the Laplace transform.

_______. "Pierre-Simon Marquis de Laplace." In *Dictionary of Scientific Biography*. Vol. 15. New York: Charles Scribner's Sons, 1978. This chronological survey of Laplace's scientific career combines discussion of significant concepts with summaries of important mathematical derivations.

Hahn, Roger. *Laplace as a Newtonian Scientist*. Los Angeles: Williams Andrew Clark Memorial Library, 1967. This short essay describes the philosophical debate concerning the status of laws of nature that occurred during Laplace's formative period at the University of Caen and his early years in Paris. Laplace's convictions about the law-governed structure of the universe are traced to his reading of d'Alembert and Marquis de Condorcet.

_______. *Pierre Simon LaPlace, 1749-1827: A Determined Scientist*. Cambridge, Mass.: Harvard University Press, 2005. Full biography of Laplace by a scholar who has studied him for decades.

Todhunter, Isaac. *A History of the Mathematical Theory of Probability from the Time of Pascal to That of Laplace*. New York: Chelsea House, 1965. Chapter 10 provides a technical and chronological account of the chief results and some of the derivations found in Laplace's publications on probability theory.

HENRIETTA SWAN LEAVITT

American astronomer

Leavitt created a brightness scale for measuring the universe, a scale called the period-luminosity relation, which, in effect, greatly expanded the size of the universe in the human imagination. She also identified more than two thousand variable stars. Her persistence in the face of gender barriers helped forge a path for women in astronomy and other sciences.

Born: July 4, 1868; Lancaster, Massachusetts
Died: December 12, 1921; Cambridge, Massachusetts
Areas of achievement: Astronomy, physics

Early Life

Henrietta Swan Leavitt (LEHV-iht) was born in Lancaster, Massachusetts, one of seven children born to the Reverend George Roswell Leavitt and Henrietta Swan Kendrick. Leavitt's natural curiosity and piety thrived under her father's Congregationalist ministry in Cambridge. Her parents had Puritan beliefs, shunning women's education in general but nonetheless encouraging their own daughter's education. Her family's move to Cleveland, Ohio, where her father accepted a ministerial appointment, placed her in proximity to Oberlin College, one of the first colleges in the United States to admit women.

At Oberlin, Leavitt studied music between 1885 and 1888, but then returned east to Massachusetts to attend the Society for the Collegiate Instruction of Women (SCIW; now Radcliffe College) in Cambridge. During her senior year at SCIW, Leavitt took an astronomy course, which sparked her interest in the discipline. She remained at SCIW so that she could take an additional astronomy course. Soon after earning her A.B. degree in 1892, Leavitt fell ill and recuperated with family in Wisconsin. This illness left her with a hearing impairment.

Life's Work

In 1895, Leavitt was appointed to a position as research assistant at the Harvard College Observatory, the first year of seven she would work there as a volunteer. She received graduate credits toward a degree in astronomy for her work at the observatory, though she never completed the requirements for the degree. Her primary assignment was establishing stars' magnitudes by referring to photographs of the universe. Her colleagues grew to admire her thoroughness, loyalty, and dutifulness.

Leavitt soon joined other astronomers at the observatory, including Annie Jump Cannon and Williamina Paton Fleming, as a permanent staff member on the authority of Edward Pickering, observatory director. One of his objectives for the observatory was "to establish a standard photographic sequence based on stars near the north celestial pole." Pickering's policy on hiring women to work in his observatory followed examples set by organizers of other professions, notably, Melville Dewey. Pickering employed highly educated women because he could pay them less than he could similarly schooled men for equivalent work. Large amounts of astronomical data were being generated by scientists at the observatory. Initially he hired men to sift through the data, but he soon discovered that the men lacked an attention to detail the work required. Pickering relied on women to analyze this data,

> ### The Period-Luminosity Scale
>
> *During her tenure at Harvard College Observatory, Henrietta Swan Leavitt observed and photographed nearly 2,500 variable stars, measuring their luminosities over time. She noted the following in a circular dated March 3, 1912.*
>
> The measurement and discussion of these objects present problems of unusual difficulty, on account of the large area covered by the two regions, the extremely crowded distribution of the stars contained in them, the faintness of the variables, and the shortness of their periods. As many of them never become brighter than the fifteenth magnitude, while very few exceed the thirteenth magnitude at maximum, long exposures are necessary, and the number of available photographs is small. The determination of absolute magnitudes for widely separated sequences of comparison stars of this degree of faintness may not be satisfactorily completed for some time to come. With the adoption of an absolute scalePeriod-luminosity scale:[Period luminosity scale] of magnitudes for stars in the North Polar Sequence, however, the way is open for such a determination.

thus freeing his scientists to gather more. Leavitt administered the observatory's photographic photometry department, earning thirty cents an hour.

By 1907, Leavitt's mastery of the tedium assigned her caught Pickering's notice. He directed Leavitt's energies toward instituting a reference for star brightness. Selecting such a standard would simplify the photographic process in the field of astronomy. Though an old science, astronomy was revolutionized by the development of the spectroscope and of photography in the nineteenth century. Leavitt, now managing the observatory's work on star brightness, solved a major problem. On film, a star was imaged, but without its actual brightness. As a remedy, Leavitt compared stars to one another to determine the measure of their respective brightness.

Leavitt worked with more than three hundred photographs from more than a dozen observatories to complete the project, and she was responsible for ensuring that telescopes were accurately corrected. This ranking of stars' brightness became known as the Harvard standard. Though it was her first immersion in a scientific project of this scale—and she soon solved additional research problems—it was her comparison of star brightness that continued to fascinate and sustain her throughout her lifetime.

Leavitt's work on the period-luminosity relationship between stars stemmed from her previous work with variable stars, those stars that change brightness. Her studies of the Cepheid variable stars in the Small Magellanic Cloud, a dwarf galaxy, allowed her to focus on their luminosity instead of their brightness or proximity to one another. She observed that brighter variables have longer periods of change and a regularity to those changes. Her ability to formulate the relationship between the brightness of stars and their periods of regularity led astronomers to calculate distances of other stars from Earth in other sections of the Milky Way Galaxy. Before Leavitt's discovery, distances between stars could be measured only within one hundred light years. After her observations, cosmic distances could be established up to ten million light years. Her discovery was crucial to the work of astronomers Ejnar Hertsprung and Harlow Shapley, and her results were published in the *Harvard College Observatory* circular of March 3, 1912, but under Pickering's name. The next year her standards were embraced by the International Committee on Photographic Magnitudes.

Female colleagues Margaret Harwood, Cannon, and Fleming recognized Leavitt's brilliance and accomplishments, as did professional organizations. She was a member of the American Association for the Advancement of Science and the Astronomical and Astrophysical Society of America. She had an honorary membership in the American Association of Variable Star Observers. Additionally, she was a member of Phi Beta Kappa and the American Association of University Women.

Leavitt never married, nor did she have children. After her father's death in 1911 she lived with her mother in Cambridge, residing there until her own death from stomach cancer on December 12, 1921. Her work on a new photographic magnitude scale was unfinished. Four years after her death she was nominated for a Nobel Prize by the Swedish Academy of Sciences.

Significance

Leavitt worked as a woman in a male-dominated field, and she worked with Pickering, who limited her tasks to those narrowly focused on tedious practical applications in lieu of theoretical deliberation. These restrictions, however, did not stop her from excelling in her field. She developed a method in which astronomers could more accurately calculate extragalactic distances, in effect expanding, in the process, the size of the universe. Had she not been stymied by Pickering and by a sexist work culture, there is no doubt that Leavitt could have added to her list of critical contributions to astronomy.

Rebecca Tolley-Stokes

Further Reading

Bailey, Solon Irving. *The History and Work of the Harvard Observatory, 1839 to 1927: An Outline of the Origin, Development, and Researches of the Astronomical Observatory of Harvard College Together with Brief Biographies of Its Leading Members*. New York: McGraw-Hill, 1931. Gives an overview of the history of the Harvard Observatory as well as brief biographies of astronomers and administrators who worked there.

Hoffleit, Dorrit. *Women in the History of Variable Star Astronomy*. Cambridge, Mass.: American Association of Variable Star Observers, 1993. Chronicles the important role that women have played and continue to play in the field of variable star astronomy.

Johnson, George. *Miss Leavitt's Stars: The Untold Story of the Woman Who Discovered How to Measure the Universe*. New York: W. W. Norton, 2005. The only full-length biography of Leavitt, placing great emphasis on her discovery of extragalactic measurement.

Jones, Bessie Zaban, and Lyle Gifford Boyd. *The Harvard College Observatory: The First Four Directorships*. Cambridge, Mass.: Belknap Press, 1971. Chronicles the Harvard College Observatory's direction during the time of Leavitt's tenure there.

Kass-Simon, G., and Patricia Farnes. *Women of Science: Righting the Record*. Bloomington: Indiana University Press, 1990. A standard reference in the field, this book contains footnoted essays on women's contributions to the sciences, including astronomy.

GEORGES LEMAÎTRE

Belgian physicist and astronomer

Building on evidence and theories adduced by several physicists, mathematicians, and cosmologists whose work, like his own, focused on the provenance and character of the universe, Lemaître, through brilliant synthesis, formulated what has become the generally accepted scientific explanation of the universe's origin, namely, the big bang theory.

Born: July 17, 1894; Charleroi, Belgium
Died: June 20, 1966; Louvain, Belgium
Also known as: Georges Abbé Lemaître (full name); Georges-Henri Lemaître
Areas of achievement: Astronomy, physics, mathematics

Early Life

Little is known of the early life of Georges Lemaître (jawrzh leh-meht-reh). He was born in the commercial, mining, and industrial center of Charleroi, Belgium, where as a youth, he commenced studying civil engineering at the local technical university. Eruption of World War I ended his engineering studies, and from 1914 until 1918 he served as an artillery officer with the remnants of the Belgian army holding the extreme left flank of the Allied armies on the western front. His experiences there led him in the war's aftermath to enter the Mechelen (Belgium) Seminary in preparation for a Roman Catholic priesthood, his ordination following in 1923. In addition to learning his priestly vocation during his years as a seminarian, Lemaître, in 1920, procured a doctorate in mathematics and physics—fields of scientific inquiry immensely stimulated by the implications of Albert Einstein's theory of relativity and its relation to regnant Newtonian physics. Already having familiarized himself with Einstein's general theory of relativity while studying at Mechelen, Lemaître, among many other aspiring scientists, was particularly intrigued by Einstein's cosmology, that is, his suggestion that the totality of the universe is fixed in its dimensions, that it is not expanding, though indeed it is full of motion and is undergoing rapid evolution.

Further pursuing the cosmological implications of Einstein's work, Lemaître, encouraged by his superiors, spent 1923 and 1924 in England at Cambridge University's solar physics laboratory. There he worked under the aegis of Sir Arthur Stanley Eddington. Eddington had contributed significantly to an understanding of the internal composition of stars and to knowledge of their motion and evolution. Equally important for Lemaître's specific curiosities, Eddington also ranked as an early and foremost champion of Einstein's general theory of relativity. Maintaining his contacts with Eddington, Lemaître subsequently moved on, further to satisfy his astrophysical inquiries, to the United States from 1925 until early 1927.

This too proved an exceedingly fruitful experience, for he was able to work both at the Massachusetts Institute of Technology (MIT) and at the famed Harvard Observatory. Between MIT and Harvard, he became acquainted with two of the world's most prominent astronomers, Harlow Shapley and Edwin Powell Hubble, each of whom was engaged variously in unraveling the structure of the universe. Lemaître, toward the end of his American studies, met Hubble during a notable conference in Washington, D.C., at which he announced his discovery of cepheids (pulsating stars) in the Andromeda nebula. Thus, having enjoyed four years abroad in touch with the world's foremost cosmologists, astronomers, and physicists, Lemaître returned to a professorship in astrophysics at the University of Louvain (Belgium) and the formal launching of his own scientific career.

Life's Work

For scientists such as Lemaître who were obsessed with questions about the structure and origins of the universe, the 1920's was a propitious time, for both the available theories and instrumentation were undergoing significant advances. Hubble's discoveries, for example, owed much to the new one-hundred-inch telescope. Similarly, the appearance of Einstein's general theory of relativity in 1917 dramatically raised the level of cosmological inquiry by proposing a finite yet unbounded universe—a theory in turn suggesting lacunae in Newtonian physics and cosmology.

However, Einstein's "static" model of the universe, as good theory should, not only sketched fresh lines of inquiry but also revealed discrepancies of its own. How, for example, could the definite "flight" of extragalactic nebulae—confirmed by astronomers' detection of the "redshift" characterizing their light signatures—be explained? Evidently these nebulae were rapidly moving away from our solar system and often from one another, in which case the universe apparently was expanding: infinitely perhaps. Confronted with such novel and mounting evidence, Newton's mechanics, with their presumed universal applicability, as well as Einstein's general theory, invaluable as they were, prompted development of other cosmological models or hypotheses. Thus, although his theory remained unnoticed for several years, Alexander Freedman, a brilliant Russian mathematician, in 1922 had modeled a relativistic universe, which, while expanding, was also curving back on itself. Similarly, the Dutch astronomer and devotee of Einstein's general theory, Willem de Sitter, whose work directly engaged Lemaître's attention, tried to demonstrate in 1917 that Einstein's universe, while apparently static, was expanding—if certain calculations were ignored. Add to these theories those of Eddington and Shapley, among others, and the intense focus of astronomers, astrophysicists, mathematicians, and cosmologists becomes apparent. This spate of new hypotheses—Lemaître's included—was not surprising, for Einstein was shortly to acknowledge that his proposal of a static universe and of a cosmological constant (which kept it static) was his worst mistake. General relativity, in sum, failed to account adequately for the flight of extragalactic nebulae, for the redshift, or for complex aspects of magnetism as it operated through vast reaches of space.

It was in this fecund intellectual environment that Lemaître, between 1925 and 1927, began advancing his own hypothesis. In a note published in the *Journal of Mathematics and Physics* in 1925, he specifically rejected de Sitter's hypothesis and two years later published his report on a universe of constant mass and variable radius and the problem confronting this perspective in view of the radial speed of extragalactic nebulae. Essentially, the position he was assuming in his depiction of expanding space, intellectually, lay somewhere between that of Einstein and that of de Sitter. He was proposing a mathematically sustainable law that linked the speed evinced by the "flight" of extragalactic nebulae with their distance from the earth. In 1929, Hubble's discovery (known as Hubble's law) that the more distant galaxies were, the faster they were receding from the earth, furnished confirmation of Lemaître's concept of an expanding universe.

Conceptualization of an expanding universe provided Lemaître with a matrix for the formulation and advancement of the hypothesis for which he is most noted, namely, that the universe, and the behavior of matter within it, had been produced by the explosion (subsequently dubbed the big bang) of what he called the "cosmic egg," or the "primitive atom." To accomplish his quantum conceptual shift from the cosmological macrocosm to the atomic microcosm meant that his inquiries were turned primarily from astronomy toward classic physics as well as to the latest developments in that burgeoning field.

From classic physics, Lemaître had to draw on that part of the law of thermodynamics linking the decay of matter to increasing divisions of energy; drawing on then recent discoveries in physics, he mathematically probed the characteristics and behavior of light, of cosmic and radioactive rays, as well as the immense releases of energy from disintegrating uranium. Boldly extrapolating both substantive and theoretical intelligence about known characteristics of the universe and those of the heaviest atoms, he proposed that, except in terms of scale, correlations existed between "exploding" or rapidly compacting and disintegrating atoms (or their side effects) and behavior evidenced by the universe and matter within it. To Lemaître, ultrapenetrating rays from space represented the preserved testimony, billions of years old, of the cosmos's primary activity. Like the earth's geological fossil record, these "fossil rays" were signatures of the age of radioactivity that preceded formation of the stars. That age itself resulted, according to Lemaître's calculations, from unimaginably immense concentrations of matter and energy—the world reduced to a single, if gigantic, particle, the "primitive atom"—which, like "observed" stars, exploded, giving birth to the universe.

Lemaître's hypothesis received its first full exposure in 1933 with his *Discussion sur l'évolution de l'univers* (discussion on the evolution of the universe) and *L'Hypothèse de l'atome primitif: Essai de cosmogonie* (1946; *The Primeval Atom: An Essay on Cosmogony*, 1950). Both works are succinct, pioneering presentations of immense and diverse learning ranging through philosophy, physics and astrophysics, astronomy, and mathematics, which left a modern, evidentially substantial, big bang theory in tolerable congruence with Einsteinian physics. Lemaître's universe, a relativistic one, was finite yet simultaneously unbounded, there being no limits to the motion of matter—whether stars, galaxies, comets, or light—within its curved space.

By the time of Lemaître's death in Louvain, Belgium, on June 20, 1966, his hypothesis, while withstanding several competitors, inevitably had undergone modifications. Indeed, by 1948, the work of George Gamow and two colleagues reversed important aspects of Lemaître's hypothesis. Lemaître, nevertheless, had propounded a theory from which others profitably had taken their own points of departure.

Significance

Cognizant of the wide range of cosmologies propounded in the past, not only those of biblical origin but also those of philosophers and naturalists such as Immanuel Kant, Comte de Buffon, and Pierre-Simon Laplace, Lemaître had a fine sense of the scientific context into which his own hypothesis fit. Like his contemporary cosmologists (astrophysicists, physicists, astronomers, mathematicians, and philosophers), he had inherited a three-dimensional universe, which, thanks to Einstein, had suddenly become four-dimensional. The addition of time, an autonomous and continuous factor, to the cosmological equation was not the sole new complexity to be grappled with. The four-dimensional universe, thanks to discoveries by the twentieth century's new breed of relativist physicists, was fleshed out with puzzling forms of mass and energy. New forms of attraction and repulsion had to be embodied in their calculations. Astronomers, as the redshift clearly indicated, had to reckon with galaxies billions of light years from the earth that were receding at fantastic speeds. Atoms themselves were found to be subdivided into subparticles of their own, each with its discrete or peculiar characteristics. The operation of gravitation had to be unraveled and recalculated. Classic speculations and Newtonian mathematical descriptions of the universe's structure were rapidly and profoundly altering.

Such were some of the baffling, yet immensely exciting, factors that encouraged, if they did not demand, fresh cosmological theorizing. This theorizing was readily forthcoming, though understandably wanting in some respects. Even Einstein was unable to sketch satisfactorily the lineaments of the cosmos. De Sitter's equations proved far less satisfactory, attaining accuracy only if one dealt with space that was empty—devoid of any matter whatsoever. Into this theoretical penumbra, Lemaître inserted his bold, rather elegant schema concerning the cataclysmic origins of the universe, which for years of intense scientific focus on the cosmos served as a seminal influence on others.

Clifton K. Yearley

Further Reading

Einstein, Albert. *Relativity: The Special and the General Theory*. Translated by Robert W. Lawson. 1920. New ed. New York: Penguin Books, 2006. Einstein's essay, "Considerations on the Universe as a Whole," should dispel the notion that Einstein is unreadable, despite the complexities of other areas of the book. This portion of one of the classic works of science is clear and straightforward. It is superb for understanding the background against which Lemaître's hypothesis was developed. There are some notes as well as modest bibliographical references and a brief index.

Farrell, John. *The Day Without Yesterday: Lemaître, Einstein, and the Birth of Modern Cosmology*. New York: Thunder's Mouth Press, 2005. Focuses on Lemaître's seminal role in postulating the big bang theory and the existence of black holes.

Gamow, George. "Modern Cosmology." *Scientific American* 190 (March, 1954): 55-63. A superb summation, clearly written and authoritative, by the Russian-born American scientist whose theory supplanted that of Lemaître. Contains illustrations, photographs, and brief bibliographical suggestions.

Hubble, Edwin P. *The Realm of the Nebulae*. New Haven, Conn.: Yale University Press, 1936. A delightful, authoritative exposition. An elegant description of the intellectual environment in which Lemaître developed his views by a figure who influenced him. Contains a bibliography and an index.

Munitz, Milton, ed. *Theories of the Universe: From Babylonian Myth to Modern Science*. New York: Free Press, 1965. A splendid collection of essays. The editor's introductory overview is excellent. Contains illustrations, a brief but good bibliography, and an extensive index.

"Obituary: Georges Lemaître." *Physics Today* 19 (September, 1966): 119. This brief obituary is actually quite revealing about the facts of Lemaitre's life.

Sagan, Carl. *Cosmos*. 1980. New ed. New York: Wings Books, 1995. A best seller that is a spinoff of the television series of the same name, by a distinguished and prolific scholar. There is no mention of Lemaître, though many other major cosmologists and their theories are mentioned. A delightful and instructive work, containing color photographs, graphs, illustrations, two appendixes, fine suggested readings, and a very useful and extensive index.

Singh, Simon. *Big Bang: The Origin of the Universe*. New York: Fourth Estate, 2004. This history of the development of the big bang theory includes discussion of Lemaître's contributions.

MARIA MITCHELL

American astronomer

The first woman in the United States to become a professional astronomer, Mitchell discovered a comet in 1847 and became the most famous American female scientist of her time, as well as a leading voice for educating women in the sciences.

Born: August 1, 1818; Nantucket, Massachusetts
Died: June 28, 1889; Lynn, Massachusetts
Area of achievement: Astronomy

Early Life

Maria Mitchell was born into a Quaker seafaring community that encouraged women to become self-reliant and well educated. Her mother and grandmother were both teachers. Her father, William Mitchell, was also a teacher who had a part-time position with the U.S. Coast Survey, for which he observed stars to check the accuracy of the chronometers that were used to tell time and chart longitude on whaling ships. From the age of twelve, Maria counted seconds for her father. She attended the academy of Cyrus Peirce, served as her father's teaching assistant at the age of sixteen, and opened her own school at the age of seventeen.

In 1836, Maria closed her school to accept the job of librarian for the Nantucket Athenaeum, which had just been founded. She taught herself mathematics, astronomy, German, and Latin by reading the books from the collection. These included textbooks by Charles Hutton and Nathaniel Bowditch and treatises by Joseph-Louis Lagrange, Pierre-Simon Laplace, and Karl Friedrich Gauss. At night, she and her father continued to study the sky, making observations with a circle and transit lent by West Point and a four-inch equatorial telescope from the Coast Survey. William Mitchell was by that time a clerk for Pacific Bank; his employers provided living quarters in which he built an observatory.

Life's Work

During a dinner party on October 1, 1847, Maria Mitchell slipped away to use the family telescope and noticed a previously unknown comet at a position five degrees above the North Star. She and her father monitored the comet for several days, computed its orbit, and shared the discovery with William Bond, the director of the Harvard Observatory. Bond then communicated the Mitchells' observations and calculations to the European scientific community. Their findings were published in England's *Monthly Notices of the Royal Astronomical Society*, and King Frederick VII of Denmark awarded to Maria the gold medal that he had offered to the first person to discover a comet with a telescope. In 1848, Maria Mitchell was elected the first female honorary member of the American Academy of Arts and Sciences, and the Seneca Falls Woman's Rights Convention cited her achievement.

Mitchell's discovery made her one of the most famous astronomers in the United States, and she was invited to join the Coast Survey in 1849. She computed the position of Venus for the *American Ephemeris and Nautical Almanac* and assisted with a survey of the Maine coast in 1852. Meanwhile, she was elected to the American Association for the Advancement of Science in 1850, and she wrote scientific articles and began to travel outside Nantucket more frequently. She continued to work at the

Athenaeum even as her fame grew but finally quit in 1856 to chaperon a wealthy girl on tours of the American South and Europe. After the girl returned home to the United States, Mitchell remained in Europe to meet other astronomers. During that trip, she became the first woman admitted to the Vatican Observatory; however, she was allowed to enter the observatory only during the daytime.

When Mitchell returned to the United States, a group that called itself "Women of America" presented her with a five-inch equatorial telescope, built by Alvan Clark, that permitted her to make more precise observations. She continued to work as a computer and field researcher for the *Nautical Almanac*. After her mother died, she moved to Lynn, Massachusetts, with her father and established a small observatory. By this time, she had become a nominal Unitarian in her religious beliefs, but she maintained Quaker values of simplicity and abolitionism. Because of her opposition to slavery, she boycotted southern cotton and always wore silk clothing.

As the U.S. Civil War came to an end in 1865, Mitchell accepted an invitation to become professor of astronomy and director of the observatory a new women's college that was being founded by Matthew Vassar. The twelve-inch telescope at Vassar College was the third-largest in the United States. Mitchell never married, so her father again accompanied her to the college campus at Poughkeepsie, New York.

At Vassar, Mitchell became known as a demanding but respected instructor who treated her students as intellectual equals and involved them directly in scientific work. For example, she and her students recorded the passage of four thousand meteors during a meteor shower in 1868. Mitchell also wrote papers on double stars, the daily photography of sunspots, an 1878 solar eclipse, and the satellites of Jupiter and Saturn that originated in part from her teaching practices. She also successfully challenged the college on the level of her salary, which was half that of younger and less-accomplished male professors. In 1869, she was elected to the American Philosophical Society.

In 1873, when Mitchell visited Europe a second time, she met with proponents of higher education for women and was ready to campaign nationally for women's rights. Afterward, she became a founder and the president (1875-1877) of the American Association for the Advancement of Women (AAW). She encouraged women to join AAW and to present scientific papers at its annual meetings. She believed both that women benefited from scientific education and that science needed women's special skills in detailed observations and teaching. From 1876 to 1888, Mitchell chaired the AAW's science committee, which prepared yearly reports on women in science. She also conducted open-ended surveys to determine where opportunities in science for women actually existed in 1876 and 1880. On June 28, 1889, she died in Lynn, Massachusetts, toward the end of her seventy-first year.

Maria Mitchell. (Library of Congress)

Significance

As an astronomer and educator, Maria Mitchell played a seminal role for women in the history of science. She repeatedly proved by example that women could contribute new knowledge and observations. She was the most famous American female scientist of the nineteenth century. She received honorary degrees from Indiana University (1853), Rutgers Female College (1870), and Columbia University (1887). Mitchell always sought to bring other women along with her. For example, her students included the academic and scientific leaders Mary Whitney, Christine Ladd-Franklin, and Ellen Swallow Richards. She promoted collaborative approaches to science in her teaching and involved her students in important research. She spoke out publicly for women and helped create a place for them in which to raise their voices through the AAW.

Despite the honors she accumulated, Mitchell always saw herself as a data collector rather than as some kind of genius. Although she hoped women would hold academic positions as professionals, she expected that most of her students would remain amateurs. She was careful never to challenge men directly in her speeches. Meanwhile, the importance of the AAW as an advocate for scientific education for women ran its course during the 1880's, as other, specialized women's organizations emerged. The AAW then shifted its emphasis to philanthropy and eventually became the General Federation of Women's Clubs.

Mitchell's significance was symbolic as well as concrete. She herself was well aware that because few women were in the scientific community, those who wished to succeed were expected to demonstrate exceptional abilities. Her willingness to talk about that double standard inspired later generations of women, who founded the Maria Mitchell Association in 1902 to preserve Mitchell's Nantucket home. In 1908, an observatory was built at her home where student research is still carried out under the auspices of Harvard astronomers.

Amy Ackerberg-Hastings

Further Reading

Albers, Henry, ed. *Maria Mitchell: A Life in Journals and Letters*. Clinton Corners, N.Y.: College Avenue Press, 2001. A well-documented account by an emeritus Vassar professor that is especially strong at conveying Mitchell's daily routine and intellectual energy.

Kendall, Phebe Mitchell. *Maria Mitchell: Life, Letters, and Journals*. Boston: Lee and Shepard, 1896. A mostly accurate account by Mitchell's sister, who destroyed several of Mitchell's papers.

Kidwell, Peggy. "Three Women of American Astronomy." *American Scientist* 78 (1990): 244-251. Describes the scientific contributions of Mitchell, Annie Jump Cannon, and Cecilia Payne-Gaposchkin and discusses what each of these women said about the role of women in astronomy.

Kohlstedt, Sally. "Maria Mitchell and the Advancement of Women in Science." In *Uneasy Careers and Intimate Lives: Women in Science*, edited by Prina G. Abir-Am and Dorinda Outram. New Brunswick, N.J.: Rutgers University Press, 1987. Emphasizes Mitchell's teaching techniques and advocacy for women's education. Concludes that Mitchell employed deliberate strategies for a changing but cautious world.

Oles, Carole. *Night Watchers: Inventions on the Life of Maria Mitchell*. Cambridge, Mass.: Alice James Books, 1985. Poems inspired by and drawn from Oles's study of Mitchell's manuscripts. Provides a literary perspective on Mitchell's significance and personal idiosyncracies.

Wright, Helen. *Sweeper in the Sky: The Life of Maria Mitchell*. Reprint. Clinton Corners, N.Y.: Attic Studio Press, 1997. A popular biography by a Vassar alumna and astronomer that carefully reconstructs Mitchell's childhood.

NABU-RIMANNI

Babylonian astronomer and mathematician

Nabu-rimanni copied and preserved astronomical tables for the computation of lunar, solar, and planetary phenomena. These accurate numerical parameters for the prediction of astronomical phenomena furthered the development and success of Greek spherical astronomy, developed to its fullest in the Ptolemaic system.

Born: Early first century B.C.E.; probably Babylonia (now in Iraq)
Died: Late first century B.C.E.; probably Babylonia
Also known as: Nuburianos; Naburiannuos; Naburiannu; Naburimannu; Naburimani
Area of achievement: Astronomy

Early Life

Because the nature of Babylonian sources is such that authors and authorship remain obscure, it is not possible to reconstruct for Nabu-rimanni (nah-BEW rih-MAHN-nee) a biography in the strict sense. The historical period with which he is associated, however, may be sketched. While Nabu-rimanni flourished in the mid-first century B.C.E., one may also define this period more broadly, that is, from roughly 300 B.C.E. to the beginning of the common era, as the Hellenistic period. As a result of the spread of Greek political influence across the Near Eastern territories once belonging to the empires of Persia, Babylonia, and Assyria, this era produced notable cultural and intellectual change from the cities of mainland Greece to Egypt in the south and Mesopotamia in the east. In particular, science flowered in Hellenistic intellectual centers such as Alexandria, and Greek astronomy, which had already begun in the fifth century, reached its height during the Hellenistic period, in part as a result of the transmission of astronomical knowledge from Mesopotamia.

Life's Work

The Hellenistic authors Strabo (64 or 63 B.C.E.-c. 25 C.E.) and Pliny the Elder (23-79 C.E.), who traveled and produced encyclopedic compendia of the knowledge and customs of the day, mention Babylonian astronomical "schools" and a few of the Babylonian astronomers by name. Thus, one finds in book 16 of Strabo's *Geographia* (c. 7 B.C.E.; *Geography*, 1917-1933) the name Naburianos (Nabu-rimmani), and also Kidenas, Sudines, and Seleucus, all associated with Babylonian cities such as Babylon, Uruk, and possibly Borsippa. Nabu-rimanni's particular or distinguishing role in the history of Babylonian astronomy, however, cannot be determined either from the Greek account or from Babylonian cuneiform sources.

Cuneiform texts yield information about the scribes only in the colophons at the ends of the inscriptions. These colophons, when complete, note the names of the owner of the document and the scribe who copied it as well as the date the tablet was written and who was king at the time. Nabu-rimanni's name is preserved on the colophon of an astronomical tablet from Babylon. The only fact, therefore, that can be established about him from Babylonian sources is that he was a scribe who copied, or possibly computed, a table of dates and positions in the sky of new and full moons for the year 49-48 B.C.E.

This particular colophon is the source for the claim that Nabu-rimanni was the inventor of the method of astronomical computation represented in his tablet. Far from showing this scribe as an innovator of Babylonian astronomy, however, Nabu-rimanni's tablet is one of the youngest of Babylonian lunar ephemerides; the oldest such tablets stem from the third century. Nabu-rimanni can therefore be credited only with preserving the tradition of Babylonian astronomy, not inventing it.

Because Nabu-rimanni is associated with a particular method of astronomical computation, but his individual contribution cannot be determined from the sources, the focus of any examination of his life's work must be Babylonian astronomy itself. Babylonian mathematical astronomy of the last three centuries B.C.E. is known from only two identifiable archives, one found in Babylon and the other in Uruk. The bulk of the texts are lunar or planetary ephemerides, which are supplemented by a smaller group of procedure texts outlining the steps necessary to generate ephemeris tables. The ephemeris tables contain parallel columns of numbers in specific sequences that represent occurrences of characteristic lunar and planetary phenomena. Each column represents a different periodic phenomenon—for example, new moons, eclipses, first visibilities, stationary points. The consecutive entries in each column correspond to dates, usually months. In the case of the ephemeris for the moon, the objective is to predict the evening of the first visibility of the lunar crescent. The appearance of the new moon defines the beginning of the month in a strictly lunar calendar. Indeed, the control of the calendar seems to have provided a major motivation for the development of mathematical methods for predicting astronomical phenomena.

For calculations, the Babylonians used a number system of base 60; that is, numbers are represented with special digits from 1 to 59, while 60, or any power of 60, is represented by 1. These "sexagesimal" numbers were written using a place-value notation system similar to decimal notation, so that for each place a digit is moved to the left, the value is multiplied by 60. The positional system was extended for fractions, which were expressed by moving digits to the right of the "ones" place, thereby dividing each time by 60. The Babylonian sexagesimal system is still preserved in the counting of time by hours, minutes (1 hour equals 60 minutes), and seconds (1 minute equals 60 seconds).

Babylonian lunar and planetary theory comprised two separate but coexistent systems, designated A and B, which are defined according to two different arithmetical methods of describing the distance covered each month by the sun. In this way, the velocity of the sun could be measured in terms of the progress of the sun in longitude, or degrees along its apparent path through the stars, the ecliptic. In system A, the progress of the sun along the ecliptic is described as being 30° per (mean synodic) month for one part of the zodiac (from Virgo 13° to Pisces 27°) and 28° 7¢30″ for the other part (from Pisces 27° to Virgo 13°). A mathematical model is thereby created whereby the sun moves with two separate constant velocities on two arcs of the ecliptic. If the sun's velocity, reckoned in terms of progress in longitude (degrees along the ecliptic), is plotted against time, the resulting graph represents a "step function." System A also implies a certain length of the solar year, namely 1 year equals 12;22,8 months, expressed sexagesimally. (Sexagesimal numbers are represented in modern notation with a semicolon separating integers from fractions.) This method and its complementary system B were both used during the period from c. 250 B.C.E. to c. 50 B.C.E.

System B assumes the motion of the sun to increase and decrease its speed steadily from month to month. The variation in velocity is bounded by a minimum and a maximum value, and within this range of velocities, the monthly change is always by a constant amount. Plotting the progress of the sun by this model produces a graph representing a "linear zig-zag function." The name Kidenas, mentioned by Strabo, Pliny, and Vettius Valens (second century C.E.), may be associated with System B, as a scribe by the name of Ki-di-nu (Kidinnu) is known from colophons of system B-type ephemerides for the years 104-101 B.C.E. The Greeks credited him with derivation of the relation that 251 synodic months equals 269 anomalistic months. This numerical relation is in fact seen in system B computations.

Systems A and B constitute theoretical mathematical models of the motion of the sun that account for the varying lengths of the seasons of the year. By analogy with solar motion, the methods of systems A and B were applied to many celestial phenomena of a cyclic character. A Babylonian lunar table deals with the determination of conjunctions and oppositions of sun and moon, first and last visibilities, and eclipses, all of which are cyclic phenomena. Planetary tables for the planets Jupiter, Venus, Saturn, Mercury, and Mars predict the dates and positions in the zodiac of the cyclic appearances of planets, such as first visibilities, oppositions, stationary points, and last visibilities. The fact that each phenomenon had its own period enabled the Babylonians to compute them independently. No general theory of planetary and lunar motion was needed, as the strictly arithmetical methods of the two systems were sufficient to predict the individual appearances of the heavenly bodies.

The goal of Babylonian astronomy was, therefore, to predict when the moon or planets would be visible. In contrast, the Greeks' goal was to develop a single model that would serve to describe and account for the motion of celestial bodies in a general sense, and from which the individual appearances of celestial bodies would follow as a consequence. The achievement of this goal was found in geometrical methods and kinematic models (explaining motion), developed by Apollonius of Perga (third century B.C.E.) and perfected by Ptolemy in *Mathematike suntaxis* (c. 150 C.E.; *Almagest*, 1948). Such geometrical concepts are not found in Babylonian astronomy.

The question of the identity of the Babylonian astronomers is not answered by the cuneiform astronomical texts. One scribe is hardly distinguished from another when the extent of one's information is the appearance of the scribe's name in a text colophon. Nevertheless, the question as to the significance of the scribes Nabu-rimanni and Kidinnu, whose names are remembered by later Greek and Roman authors, remains. The belief that they were the inventors of the systems A and B gains no support from the cuneiform texts. Indeed, the establishment of dates for the invention of systems A and B has proved difficult; thus, statements concerning the origins of Babylonian mathematical astronomy, in regard to both chronology and the role of individual scribes, must for the time being remain inconclusive.

Significance

Babylonian lunar and planetary theory became the foundation for the development and further refinement of astronomy by the Greek astronomers Apollonius, Hipparchus, and Ptolemy. In very general terms, one can enumerate the various Babylonian contributions to Greek astronomy and thereby to the development of science in general as follows.

About 300 to 200 B.C.E., the Greek astronomers adopted the Babylonian sexagesimal number system for their computations and for measuring time and angles (360 degrees in a circle with degrees divisible by minutes and seconds). The use of arithmetical methods characteristic of Babylonian astronomical tables continued, particularly in the procedures used by Hellenistic astrologers. While Greek astronomical theory depended on geometrical and kinematic models, neither of which was used in Babylonian astronomy, the parameters used in constructing those models were Babylonian. Indeed, the success of Greek astronomical theory rests on the accuracy of the parameters established by Babylonian astronomers (such as the length of the solar year given above).

The concepts related to the parameters must also have been transmitted. Such concepts were, for example, the following components of lunar motion: longitude, latitude (angular distance of the moon from the ecliptic), anomaly (irregularity in motion), and the motion from east to west of the line of nodes (the two diametrically opposed points where the moon's orbital plane intersects the plane of the ecliptic) in a nineteen-year cycle.

Also of use for Greek planetary theory were the period relations essential to the determination of successive occurrences of a periodic phenomenon in Babylonian astronomy. Through the wide acceptance of the Ptolemaic tradition, as evidenced by Indian and Islamic astronomy, the direct influence of Babylonian astronomy on the Greek world had even greater impact as the ultimate impetus for the quantitative approach to celestial phenomena.

Francesca Rochberg-Halton

Further Reading

Aaboe, Asger. "Observation and Theory in Babylonian Astronomy." *Centaurus* 24 (1980): 14-35. Suggests a reconstruction of the process by which the Babylonian systems of computing astronomical phenomena were developed.

_______. "Scientific Astronomy in Antiquity." In *The Place of Astronomy in the Ancient World*, edited by F. R. Hodson. London: Oxford University Press, 1974. Summary and explanation of the mathematical content of the Babylonian astronomical ephemerides and discussion of what constitutes a "scientific" astronomy.

Hunger, Hermann, and David Pingree. *Astral Sciences in Mesopotamia*. New York: Brill, 1999. This account of the beginnings of astronomy and astrology in the ancient Near East discusses the place a given tablet has in the development of astronomy both within Mesopotamian culture and outside it.

Neugebauer, Otto, ed. *Astronomical Cuneiform Texts*. New York: Springer Verlag, 1983. A transliteration and translation of the cuneiform astronomical texts, accompanied by an analysis.

_______. *The Exact Sciences in Antiquity*. 2d ed. New York: Dover, 1969. Concise history of mathematics and astronomy in Babylonia, Egypt, and Greece.

Van der Waerden, B. L. *The Birth of Astronomy*. Vol. 2 in *Science Awakening*. 4th ed. Princeton, N.J.: Scholar's Bookshelf, 1988. More accessible and less technical but also less reliable than Neugebauer. Gives an overall account of the phenomena of interest in Babylonian mathematical as well as pre-mathematical astronomy.

JAN HENDRIK OORT

Dutch astronomer

Oort was one of the most significant astronomers of the twentieth century. He was the first to postulate the vast swarm of comets known subsequently as the Oort cloud. He was one of the pioneers of radio astronomy, and he was one of the leaders in establishing the structure of the Milky Way Galaxy.

Born: April 28, 1900; Franeker, the Netherlands
Died: November 5, 1992; Leiden, the Netherlands
Areas of achievement: Astronomy, science

Early Life

Jan Hendrik Oort (awrt) was the son of Abraham Hermanus Oort, a physician, and Ruth Hannah Faber Oort. His grandfather, a professor of Hebrew at the University of Leiden, was one of the translators of the Bible into the Dutch Leiden edition. Following his graduation from the Leiden gymnasium (secondary school) in 1917, Oort enrolled in the University of Gröningen, where he studied under the prominent astronomer Jacobus Kapteyn. He spent the years 1922 to 1924 in the United States as a research assistant at Yale University and became familiar with Harvard University astronomer Harlow Shapley's study of the Galaxy. On his return to the Netherlands, Oort began working at the University of Leiden, an association that would last throughout his career. He was appointed an instructor in 1926, after receiving his Ph.D. from the University of Gröningen with a dissertation on high-velocity stars.

Oort married Johanna Maria Graadt van Roggen on May 24, 1927. In that same year, he began establishing his reputation as a major scholar with the presentation of two papers on the rotation of the galactic system.

Life's Work

Although it hardly reflects the full range of his accomplishments in the world of astronomy, Oort's name and reputation are guaranteed longevity through the Oort cloud—the conglomeration of comets he hypothesized in 1950—and, to a lesser extent, by the Oort constants of galactic rotation and Oort's limit, the value of density near the Galaxy's midpoint. Despite the public recognition inherent in the name of the comet cloud, however, Oort's greatest achievements lie in his contributions to an understanding of the structure and function of the Galaxy.

Oort began his career by studying under Jacobus Kapteyn, one of the pioneers of galactic research. By 1922, Kapteyn had spent sixteen years mapping the Galaxy. Although Kapteyn's map showed the Galaxy as only half its actual size, he was correct in his assessment of its shape as a disk with a central bulge narrowing toward the edges (something like a discus). After sharing the Bachiene Foundation Prize in 1920 for his paper on stars of the F, G, K, and M types, Oort turned his attention to Kapteyn's researches. Oort began studying stars of high velocity, which had proved difficult to reconcile with a galactic model that had the Sun near its center. His analysis of high-velocity stars and his close observation of Shapley's work during his research assistantship in the United States convinced Oort that the Galaxy was much larger than Kapteyn had assumed.

Oort's two 1927 papers furthered the work of Bertil Lindblad of the Stockholm Observatory. The papers argued that the Galaxy rotates not as a unit but more like the solar system, with those stars far from the center moving slower and those close to the center, faster. Through elaborate calculations involving rotational velocities of various stars, Oort was able to estimate the galactic mass, the number of stars, and the dimensions of the Galaxy. As Bengt Stromgren observed in awarding the Vetlesen Prize to Oort, "... it was only through Professor Oort's detailed analysis . . . that the hypothesis was finally proved. This is an example of one paper changing the whole outlook of the astronomical community."

In 1935, Oort was promoted to professor of astronomy, and, during that decade, he turned his attention to the distribution of stars in the Galaxy, realizing that Kapteyn's problem had been the interstellar dust that obscures the view of the distant parts of the Galaxy. Although Oort's theory of galactic structure was accepted by the scientific community, it was not possible to test his hypothesis until the development of radio astronomy. During World War II, with the occupation of the Netherlands by the Nazis, the Leiden Observatory was closed, and Oort and his colleagues continued their work only with the greatest of difficulties. Nevertheless, one of the most important discoveries for modern astronomy was made by Oort's colleague Hendrik van de Hulst. Van de Hulst's work was based on the principle that the hydrogen atom periodically emits radio waves at a constant wavelength of 21 centimeters. Though this characteristic is useless in local applications, in the vastness of space there is a sufficient number of atoms to be measured. Immediately following the war, Oort and van de Hulst began working toward a test of van de Hulst's hypothesis, but it was not until 1951 that they were able to do so. The theory, which had been independently reached by the Soviet scientist I. S. Shklovsky, proved to be true and, coupled with the building of large radio telescopes, led to the establishment of radio astronomy, the most important step in studying the Galaxy. The 21-centimeter waves pass through the dust and gas and allow astronomers to "see" as they cannot with optical telescopes. Meanwhile, also in collaboration with van de Hulst, Oort, who had been appointed director of the Leiden Observatory following the war, in 1946 published a study of the clouds of gas and dust and subsequently showed that these clouds surrounding bright, hot stars can provide the material for the spontaneous formation of new stars.

In 1950, Oort made public the theory that led to the naming of the Oort cloud. His observations of the actions of nineteen long-period comets led him to the conclusion that there is a vast body of comets—in the trillions—consisting largely of debris from the disintegration of a planet. As a result of gravitational forces from various stars, comets are periodically thrust into orbits that bring them into the solar system. Oort theorized that there must be a swarm of these comets extending almost to the nearest stars, but their tie to the solar system suggests that they were formed soon after the planetary system itself. It is again a measure of Oort's careful reasoning and analysis that his theory is accepted almost universally despite the total absence of visual evidence.

By 1954, Oort and van de Hulst had completed a map of the spiral structure of the Galaxy's outer region based on their radio-wave analysis. Oort also showed that the part of the Galaxy that is visible is far too small to account for its gravitational attraction on distant stars—evidence of a black hole in the Galaxy. In 1956, Oort and van de Hulst revealed the discovery of radio signals produced by ionized hydrogen in space. Further investigation showed that these emissions occurred across the Galaxy, and, using the 21-centimeter wavelength, they established the rotational period of the Galaxy at 225,000,000 years.

In 1956, with T. Walraven, Oort engaged in the first extensive examination of the light polarization in the Crab nebula, a gaseous mass assumed to be the result of the fabled supernova of the year 1054. In a 1957 article in *Scientific American*, Oort argued that the light of this nebula consists of synchrotron radiation produced by high-velocity electrons spiraling about large magnetic fields, in accordance with a theory of Shklovsky. This phenomenon is believed to be responsible for the significant radio-wave radiation from the Crab nebula and other radio sources in space.

In the late 1950's, Oort and Australian Frank Kerr discovered that the galactic core consists of "turbulent" hydrogen that is constantly expanding outward only to be replaced by other hydrogen falling back into the center from the top and bottom of the central bulge. Oort's observations suggest that a massive explosion in the core of the disk, perhaps ten million years ago, produced this turbulence. Further radio analysis allowed him to establish the Galaxy as a rapidly whirling disk with a jagged edge produced by the juxtaposition of rapidly rotating material on the inside and material of slow rotation on the outside. In 1964, Oort and G. W. Rougoor discovered a ring of neutral hydrogen 3,000 parsecs from the galactic center. These gases, which form the spiral arms of the Galaxy, are expanding rapidly outward and also provide support for the hypothesis of a major explosion.

Oort then began studying the space through which the Galaxy moves, both on its own and as part of the cluster of galaxies to which it belongs. His most significant observation was the nearly limitless clouds of hydrogen through which the Galaxy moves. He hypothesized, from the fact that the densest part of this hydrogen corona was above, that the Galaxy was moving upward. He also theorized that while some gas was absorbed and other gas was moving outward, the fact that some was moving inward would double the mass of the Galaxy every three billion years and also effect a change in its shape.

In addition to his duties in Leiden, Oort served from 1935 to 1948 as general secretary of the International Astronomical Union and from 1958 to 1961 as its president. He received numerous honorary degrees and other awards, including the 1942 Bruce Medal from the Astronomical Society of the Pacific, the Gold Medal of the Royal Astronomical Society in 1946, and the Vetlesen Prize (established to supplement the Nobel Prizes in areas where they are not awarded) in 1966.

Significance

Jan Hendrik Oort's long and distinguished career was characterized by steady progress toward the solutions to the problems that he undertook. From the beginning, he demonstrated the ability to work well with his colleagues and students. From his first modification of Kapteyn's galactic model, he demonstrated the capacity for detailed and painstaking research even—as in the case of the German occupation—in the face of exceptional difficulty. While the ever more sophisticated methods of modern astronomy have modified or enhanced some of his theoretical work, his methods were so precise, his research so painstaking, and his arguments so convincing that the scientific mainstream accepted his theories even when no direct visual evidence existed.

Highly respected by his contemporaries, Oort cast a long shadow. His galactic formulations and his pioneering work in radio astronomy, his postulation of the comet cloud and his analysis of the Crab nebula, his administrative work, and his encouragement of others justify Carl Sagan's assertion that "Jan Oort . . . perhaps more than any other person in the twentieth century has revolutionized our knowledge of the Galaxy."

Daniel J. Fuller

Further Reading

Freeman, Ken, and Geoff McNamara. *In Search of Dark Matter in the Universe*. New York: Springer, 2006. A history of the dark matter problem, from Oort's initial erroneous discovery of dark matter to contemporary explanations of it.

Pfeiffer, John. *The Changing Universe: The Story of the New Astronomy*. New York: Random House, 1956. A good popular introduction to radio astronomy. The author devotes several pages to the work of Oort and Van de Hulst.

Ritchie, David. *Comets: The Swords of Heaven*. New York: New American Library, 1985. One of the better books to capitalize on the arrival of Halley's Comet. Contains a brief discussion of the Oort cloud.

Sagan, Carl, and Ann Druyan. *Comet*. New York: Random House, 1985. In one of his best-selling scientific works, Sagan and Druyan have assembled an impressive study of comets from both a scientific and a cultural perspective. Their discussion of Oort and the Oort cloud is outstanding. Profusely illustrated, this is a truly beautiful book.

Stromgren, Bengt. "An Appreciation of Jan Hendrik Oort." In *Galaxies and the Universe*, edited by Lodewijk Woltjer. New York: Columbia University Press, 1968. The Vetlesen tribute to Oort by the eminent Danish astronomer is an excellent commentary on Oort's career to that point.

Struve, Otto. "The Origin of Comets." In *The Origin of the Solar System*, edited by Thornton Page and Lou Williams Page. New York: Macmillan, 1966. This article, which first appeared in 1950, summarizes one of Oort's papers, which concluded with the theory of the comet cloud.

Tayler, R. J. *Galaxies: Structure and Evolution*. London: Wykeham, 1978. This book in the Wykeham Science series is aimed at high school seniors and first-year undergraduates. The book presents the full mathematical details and applications of both Oort's constants and Oort's limit.

GEORG VON PEUERBACH

Austrian mathematician and astronomer

A pre-Copernican astronomer, Peuerbach accepted, reintroduced, and extended the ideas of Ptolemy and made original astronomical observations. His published works include a look at new theories of the planets, which became a standard astronomy textbook, and an aid for the calculation of eclipses.

Born: May 30, 1423; Peuerbach, Austria
Died: April 8, 1461; Vienna, Austria
Also known as: Georg von Purbach; Georg von Peurbach; Georgius Aunpekh de Pewrbach
Areas of achievement: Mathematics, astronomy

Early Life

Georg von Peuerbach (gay-awrg fawn PEWR-bahk) was born in Upper Austria. Other than his father's name, Ulrich, nothing is known about his early years. In 1446, he entered the University of Vienna with the name Georgius Aunpekh de Pewrbach, and earned a bachelor's degree in 1448. By this time, he had probably already begun the study of astronomy under the followers of John of Gmunden (d. 1442).

Between 1448 and 1453, Peuerbach traveled to several countries, including France, Germany, and Italy, and lectured in Padua and Ferrara. In Ferrara, he met the most important contemporary Italian astronomer, Giovanni Bianchini, who tried to convince him to teach at an Italian university; Peuerbach had been given offers to teach at Bologna and Padua. In his later years, Peuerbach knew Nicholas of Cusa, and it is possible that he met him in Rome during this period. Peuerbach received a master's degree from the University of Vienna in 1453. Around this time, he became the teacher and colleague of Johann Müller, known as Regiomontanus.

From 1453 to 1456, Peuerbach corresponded with the imperial astrologer to Holy Roman Emperor Frederick III, in Wiener Neustadt. With the astrologer's encouragement, the impoverished Peuerbach accepted in 1454 the position of court astrologer to King László V of Hungary, and later became the imperial astrologerAstrology;Hungary.

While Peuerbach's service to king and emperor employed his knowledge of astronomy and astrology, his contribution to the University of Vienna involved Humanistic studies. He was among the scholars who brought the revival of classical learning to the University of Vienna by lecturing on Vergil's *Aeneid* (c. 29-19 B.C.E.), on Juvenal, and possibly on Horace, and by participating in a disputation on the art of oratory and poetics in 1458. Scholars believe that Peuerbach was the author of a treatise on letter writing. He also wrote poetry in Latin.

Life's Work

Peuerbach's correspondence with Nihil reveals that during the 1450's, Peuerbach engaged in the construction of horseshoe-shaped sundials with magnetic needles, the making of almanacs—necessary for his astrological work—and the interpretation of comets. For his almanacs, Peuerbach revealed that he had mastered the use of the thirteenth century Alfonsine tables, the first set of astronomical tables in Europe, to determine planetary positions and movement and that he was learning to use the newer tables of Bianchini. Twenty works by Peuerbach are known.

With the 1456 appearance of Halley's comet, Peuerbach became the first to attempt to calculate the length of a comet's tail and a comet's diameter. His results—4 German miles (one German mile equals 25,000 feet, or 7,620 kilometers) for the diameter and 80 German miles for the tail—were erroneous, however, because he based them on the doubtful values for the diameter of Earth and the distance of the Moon computed by Ptolemy (c. 100-c. 178) and Archimedes (c. 287-212 B.C.E.). This work provided inspiration for Regiomontanus and sixteenth century astronomers to produce additional important works about comets.

Peuerbach contributed also to the measurement of space and time. Among the instruments on which he wrote papers were the astrolabe quadrant, the astrolabe, a sundial with gnomon, and the geometric square, a device to measure the heights of constellations and other distant terrestrial objects. The geometric square had rods forming a square, with 1,200 gradations on two of the rods and movable points with a sighting hole. He also invented an instrument to represent the true new moon.

Much of Peuerbach's work involved direct observations. He revised the latitude of Vienna, setting it at 48.22°, closer to the modern value of 48.12°. With Regiomontanus, he observed a lunar eclipse in 1457 and two in 1460. As calculations based on the Alfonsine tables gave an incorrect midpoint for the 1457 eclipse, Peuerbach reorganized and expanded the tables to facilitate computations greatly. Finished around 1459 and available in manuscript form, the work was edited and published in 1514 as *Tabulae eclipsium super meridiano*

Viennensi (tables of eclipses above the Viennese meridian). Astronomers throughout the sixteenth century recognized the monumental efforts put into the work. Peuerbach and Regiomontanus used the *Tabulae eclipsium super meridiano Viennensi* to determine the eclipses of 1460.

Peuerbach was the author of an extremely important textbook based on Ptolemy's second century Earth-centered astronomy, *Theoricae novae planetarum* (English translation, 1987), published by Regiomontanus in 1474. The standard text until this time had been the anonymously written *Theorica planetarum communis* (thirteenth century; theory of the planets). The topics covered in Peuerbach's text covering new theories of the planets include the Sun, Moon, superior planets, Venus, Mercury, eclipses, the theory of latitude, and the motion of the eighth sphere based on the Alfonsine tables. As an elementary work, it provided many definitions of technical terms. In this work, Peuerbach attempted to reconcile two conceptions explaining the motion of the planets: the planet-bearing crystalline spheres, which Ptolemy accepted in *Hypotheseis ton planomenon* (planetary hypotheses) and the epicycle/deferent system of Ptolemy. He presented a model of the solid spheres based on the work of the Arab mathematician and astronomer, Alhazan (965-1039). These spheres remained part of the astronomical tradition until disproved by Tycho Brahe.

Peuerbach also explained the particular motions of the planets according to Ptolemy's epicycles. It may also be pointed out the Peuerbach suggested that the Sun controlled the motions of the planets and thus anticipated the new role the Sun acquired during the scientific revolution.

Another major work was Peuerbach's abridgment of Ptolemy's *MathTmatikT suntaxis* (c. 150; better known as *Almagest*, 1952). In 1460, Cardinal Bessarion, a Greek scholar and Catholic Church official, invited Peuerbach to prepare an epitome of the *Almagest*, that is, a short, understandable translation with explanatory comments, which was called *Epitoma magesti Ptolemaei* (1496). Peuerbach knew Gerard's Latin translation almost by memory, but insisted on the aid of Regiomontanus, who knew Greek and could work with the original. Peuerbach had finished six of the thirteen books when he died at age thirty-seven, having obtained a promise from Regiomontanus to complete the translation. Regiomontanus completed it by 1463 and died in 1476. The epitome was first published in 1496 with many errors, but better editions appeared in 1543 and 1550.

Other contributions include Peuerbach's work in mathematics. He wrote a short treatise on the computation of sines and chords. In another work, he presented computations of sines at intervals of ten minutes. Following the work of Arab scholars, he recognized for astronomical calculations the superiority of using sines rather than chords, which Ptolemy had used. Finally, he produced a popular elementary mathematical book on computation using integers and fractions, a book that was printed several times in the late fifteenth and early sixteenth centuries. Peuerbach is also noted for using Arabic numerals, which were still not universally employed in the fifteenth century.

Significance

Peuerbach's work clarified Ptolemy's astronomy and helped to bring consistency to the work of Peuerbach's contemporaries. His influence continued until after Copernicus began the revolution that overturned Ptolemaic, Earth-centered astronomy.

His *Theoricae novae planetarum* and *Epitoma magesti Ptolemaei* provided contemporary scholars with definitions of technical terms employed until the early seventeenth century. *Theoricae novae planetarum* went through fifty-six editions by 1650 and was the subject of numerous commentaries.

To the time of Galileo and Kepler, scholars, including Copernicus, relied on the *Epitoma magesti Ptolemaei* as the basic treatise on Ptolemy's astronomy. Finally, astronomers used his tabulae well into the sixteenth century, when new astronomical tables were prepared.

Peuerbach represents the use of Arabic learning in the West because he worked with Arabic numerals in computing sines, and he adopted the crystalline spheres of the Alfonsine tables.

Kristen L. Zacharias

Further Reading

Aiton, E. J. "Pürbach's *Theoricae novae planetarum:* A Translation with Commentary." *Osiris* (2d series) 3 (1987): 4-43. A short description of Peuerbach's life that introduces the translation of his important theory of the planets, including drawings of planetary motions.

Dobrzycki, J., and R. L. Kremer. "Pürbach and Maragha Astronomy? The Ephemerides of Johannes Angelus and Their Implications." *Journal of the History of Astronomy* 27, no. 3 (1996): 187-237. Argues that Peuerbach was familiar with Arab astronomy and that it influenced his work.

Margolis, Howard. *It Started with Copernicus: How Turning the World Inside Out Led to the Scientific Revolution.* New York: McGraw-Hill, 2002. Chapter 1 covers pre-Copernican astronomy, based on Ptolemy, on which Peuerbach's new theory of the planets was highly influential.

North, John. *The Norton History of Astronomy and Cosmology.* New York: W. W. Norton, 1995. An excellent history of astronomy, useful again for a discussion of Peuerbach's influence.

Zinner, Ernst. *Regiomontanus: His Life and Work.* Translated by Ezra Brown. New York: North-Holland, 1990. A biography of Peuerbach's teacher, Johannes Müller, also known as Regiomontanus, which provides details about Peuerbach's life and work as well.

PTOLEMY (ASTRONOMER)

Alexandrian scientist

Ptolemy's scientific work in astronomy, mathematics, geography, and optics influenced other practitioners for almost fifteen hundred years.

Born: c. 100 C.E.; possibly Ptolemais Hermii, Egypt
Died: c. 178 C.E.; place unknown, possibly Egypt
Also known as: Claudius Ptolemaeus
Areas of achievement: Astronomy, mathematics, geography

Early Life

Very little is known about the life of Ptolemy (TOL-uh-mee). He was born in Egypt at the end of the first century C.E., but his birth date and birth place and his life thereafter are subjects of speculation. It is thought that he might have been born in the Grecian city of Ptolemais Hermii in Upper Egypt and that he might have lived to the age of seventy-eight. It has been suggested that he studied and made astronomical observations, staying for more than half of his life among the elevated terraces at the temple of Serapis in Canopus near Alexandria, where pillars were erected with the results of his astronomical discoveries engraved on them. He was probably the descendant of Greek or Hellenized ancestors and obtained Roman citizenship as a legacy from them.

Much more is known about the age in which Ptolemy lived. It was a century during which Rome ruled the Mediterranean world and during which four successive Roman emperors, Trajan, Hadrian, Antoninus Pius, and Marcus Aurelius, built roads and bridges, opened libraries and colleges, and maintained Rome's power and peace. It was a time when educated men spoke Greek as well as Latin, when Athens was still honored for its cultural traditions, when Marcus Aurelius wrote his *Ton eis heauton* (c. 171-180; *Meditations*, 1634) in Greek, and Greek was still the language of science and the arts.

Ptolemy, who probably used the libraries at Alexandria, was strongly influenced by a Greek scientist, Hipparchus (fl. 146-127 B.C.E.), who propounded the geocentric theory of the universe. As far back as the fourth century B.C.E., the leading view of the nature of the universe had the sun, moon, and planets revolving around the fixed Earth in concentric spheres. The competing theory was first advocated by Aristarchus of Samos (fl. c. 270 B.C.E.). Aristarchus discovered that the sun was much larger than Earth, and this discovery was the basis for his argument that Earth and all other planets revolved around a fixed sun and stars in circles. Yet the heliocentric theory could not be demonstrated by observable phenomena as long as it was thought that the sun was the center of a circle rather than of an ellipse. Hipparchus rejected the contention of Aristarchus, insisting on "saving the phenomena," that is, adhering to the observations. His further scientific speculations founded on the geocentric theory were the legacy to Ptolemy some two centuries later.

Life's Work

Some historians maintain that Ptolemy merely plagiarized from Hipparchus; others have said that Ptolemy superseded Hipparchus and made the work of the earlier scientist superfluous. In fact, it could be said that Ptolemy immortalized Hipparchus by acknowledging the debt he owed to his distant predecessor and by frequently quoting from him.

Whatever historical assessment is more correct, there is no doubt that Ptolemy's work in astronomy alone lasted for more than fourteen hundred years, until the great scientific achievements of Nicolaus Copernicus (1473-1543) and Johannes Kepler (1571-1630). Ptolemy used new instruments or improved on old ones to make his observations. In the *Mathematike syntaxis* (c. 150 C.E.; *Almagest*, 1948), one of his most significant books, he utilized the mathematical methods of trigonometry to prove that Earth was a sphere and went on to postulate that the heavens were also spheres and moved around an immobile Earth in the center of the solar system. He dealt with the length of the months and the year and the motion of the sun; he covered the theory of the moon; and he figured out the distance of the sun, and the order and distances of the planets, from Earth. Much of this was not new, not original; the *Almagest* was essentially a restatement of astronomical knowledge available three hundred years earlier. Yet Ptolemy was able to synthesize that scientific information into a system and to expound it in a clear and understandable manner. He was a teacher, and he taught well.

Ptolemy's contribution to mathematics was even more significant. Hipparchus had invented spherical and plane trigonometry for the use of astronomers. Ptolemy then perfected this branch of mathematics so that, unlike his astronomical system, which was finally discredited, the theorems that he and Hipparchus devised form the permanent basis of trigonometry.

The *Almagest*, in which trigonometry was utilized to measure the positions of the sun, Earth, moon, and planets, was later translated into Arabic and then Latin, and so also was Ptolemy's *Geographike hyphegesis* (second century C.E.; *The Geography of Ptolemy*, 1732). Ptolemy attempted with considerable success to place the study of geography on a scientific foundation. His book, written after the *Almagest*, was modeled after the work of Marinus of Tyre (fl. second century C.E.), but Ptolemy added a unique dimension by placing his predecessor's information into a scientific structure. He assumed that Earth was round, that its surface was divided into five parallel zones, and that there were other circles from the equator to the poles. He was the first geographer to write of "parallels of latitude" and "meridians of longitude." Ptolemy, however, did make one crucial mistake. Along with other ancient geographers,Geography; he underestimated the circumference of Earth, and as a consequence few latitudes were established correctly (and, since the means were not available, no longitudes were established).

What most attracted the interest and attention of earlier geographers and of Ptolemy was the size of the inhabited world: in the north, Thule (the present Shetland Islands); in the west, the Fortunate Islands (the Canary Islands and Madeira); and in the south and east, the vast continents of Africa and Asia. Although they overestimated the size of both the eastern and southern continents, Ptolemy's findings, and Marinus's before him, were based on new knowledge derived from travelers' accounts of the silk trade with China and from sea voyages in the Indian Ocean. Ptolemy revised some of Marinus's estimates of

Ptolemy the astronomer. (Library of Congress)

the length and breadth of Asia and Africa, extending Asia eastward and Europe westward. More than a thousand years later, Christopher Columbus (1451-1506), who relied on Ptolemy's *Geography*, was led to believe that it was possible to reach Asia by a direct route across the Atlantic Ocean.

Ptolemy's *Geography* is restricted to mathematical calculations; he did not write about the physical attributes of the countries he charted or the people who inhabited them. His tables, stating the location of places in terms of latitude and longitude, gave a false impression of precision; he made frequent errors because of his basic misestimate of the size of Earth. Still, Ptolemy's objective to draw a world map was noteworthy. His educated guess as to the location of the sources of the Nile River was remarkable, and his use of the terms "latitude" and "longitude" was a distinct contribution to the advancement of geographical knowledge.

While Ptolemy is well-known among historians of science for his volumes on astronomy and geography, it is also necessary to consider his writings on astrology, which in the ancient world was the "science" of religions. His volume *Apotelesmatika* (second century C.E.; *Ptolemy's Quadripartite: Or, Four Books Concerning the Influences of the Stars*, 1701; commonly called *Tetrabiblos*, "four books") is important partly because it was more famous than the *Almagest* and partly because it reflects the popular thinking of his age. The *Tetrabiblos* is a summary of Egyptian, Chaldean, and Greek ideas. It attributes human characteristics to the planets, such as masculine and feminine, beneficent and malevolent. It predicts the future of races, of countries and cities, and speaks of catastrophes, natural and human: wars, famine, plagues, earthquakes, and floods. It also expounds on such subjects as marriage, children, the periods of life, and the quality of death. Translated into Arabic, Latin, Spanish, and English, it influenced generations of Europeans (and, later, Americans) and formed the basis of modern astrological beliefs.

There are many historians of science who deplore the superstitions that pervade the *Tetrabiblos* and dismiss it as an unfortunate effort. The great historian George Sarton wrote, however, that "we should be indulgent to Ptolemy, who had innocently accepted the prejudices endemic in his age and could not foresee their evil consequences. . . ."

Significance

It would be unreasonable to expect great scientific breakthroughs during the second century C.E., and they did not happen. What did occur was the gradual advancement of knowledge to which Ptolemy contributed. Not only did Ptolemy write the *Almagest* and the *Geography*, adding new and significant materials to those of his predecessors, but he also attempted to illuminate the science of optics and the art of music. In the first case, although little was known about the anatomical and physiological structure of the eye, he devised a table of refraction, and his book reveals that he understood that a ray of light deviates when it passes from one medium into another of a different density. He addressed the role of light and color in vision, with various kinds of optical illusions and with reflection. Ptolemy's volume on music theory, known as the *Harmonika* (second century C.E.; *Harmonics*, 2000), covers the mathematical intervals between notes and their classification. He propounded a theory that steered a middle ground between mathematical calculations and the evidence of the ear. Observation was again a guiding principle of his art as well as his science.

Other work on mechanics, dimensions, and the elements was done but has not survived. What did survive had great influence on the Arabic science of astronomy, led to the rise of European astronomy, and influenced the work of Copernicus himself in the fifteenth century. The *Geography*, also translated into Arabic in the ninth century, was amended to describe more accurately the territories under Islamic rule; in the West, where the work became known in the fifteenth century, it was a catalyst of cartography and to the work of the Flemish cartographer Gerardus Mercator (1512-1594). Ptolemy's work on optics inspired the great improvements made by the Arabic scientist Ibn al-Haytham (d. 1039), and his work became the foundation of the *Perspectiva* of Witelo (c. 1274), the standard optical treatise of the late Middle Ages.

Just as there is no exact knowledge of Ptolemy's birth date, there is no reliable information about when and where he died and under what circumstances. Yet those biographical facts are not that important; what is significant is the scientific legacy that was transmitted through the centuries. Ptolemy was not, as one expert has argued, an "original genius"; his forte was to take existing knowledge and to shape it into clear and careful prose.

David L. Sterling

Further Reading

Barker, Andrew. *Scientific Method in Ptolemy's "Harmonics."* New York: Cambridge University Press, 2001. This study illuminates not only the *Harmonics* but also Ptolemy's observational methods in general.

Grasshoff, Gerd. *The History of Ptolemy's Star Catalog.* New York: Springer-Verlag, 1989. A careful study of the relationship between Ptolemy's work and that of Hipparchus, showing that charges of plagiarism are anachronistic and based in a misunderstanding of Ptolemy's scientific objectives.

Irby-Massie, Georgia L., and Paul T. Keyser. *Greek Science of the Hellenistic Era: A Sourcebook.* New York: Routledge, 2001. Chapters cover the main scientific disciplines in the Hellenistic era, providing a historical and scientific context for Ptolemy's work in many fields.

Ptolemy. *Harmonics: Translation and Commentary.* Translated by Jon Solomon. New York: Brill Academic Publishers, 2000.

_______. *Ptolemy's "Almagest."* Translated by G. J. Toomer. 1984. Reprint. Princeton, N.J.: Princeton University Press, 1998.

_______. *Ptolemy's "Geography."* Translated by Alexander Jones and J. Lennart Berggren. Princeton, N.J.: Princeton University Press, 2000.

_______. *Tetrabiblos.* Translated by F. R. Robbins. Loeb Library. Cambridge, Mass.: Harvard University Press, 1980. For students and readers who may want to sample the scientific works of Ptolemy. All have useful introductions and annotations.

HENRY NORRIS RUSSELL

American astronomer

Russell became one of the first astrophysical theorists in the United States by bringing theory into astronomy, which traditionally had been a descriptive science. In contrast, he sought to explain his observations and to understand the physics of stars. He is best known, however, for his work in developing the Hertzsprung-Russell diagram, a major tool used to understand the composition and evolution of stars.

Born: October 25, 1877; Oyster Bay, New York
Died: February 18, 1957; Princeton, New Jersey
Areas of achievement: Astronomy, physics

Early Life

Henry Norris Russell was born the first son of a Presbyterian pastor, Alexander Gatherer Russell, in Oyster Bay, New York. He was raised in a strict Christian home and had a religious upbringing that effected him profoundly. Russell remained deeply faithful his entire life; however, because his father was quite harsh in his discipline, the young Russell was closer to his mother, Eliza Hoxie Norris Russell. In 1890, Russell entered the Princeton Preparatory School, and stayed with his maternal aunt, Ada Louise Norris, in Princeton, New Jersey, during the school year. He would return to Oyster Bay during the summer months. Russell proved to be an excellent student.

In 1893, just one month before his sixteenth birthday, Russell entered the College of New Jersey (now Princeton University). In 1897 he graduated *insigne cum laude*, that is, with the highest honor bestowed on any student by Princeton. He continued his studies and received his Ph.D. in 1900. His doctoral work led to the development of methods to study binary star orbits. However, health problems forced Russell to interrupt his work for the rest of that year.

By 1902, Russell had traveled to Cambridge University, England, to study and do research. He returned to Princeton in 1905 to accept a faculty position, living again with his Aunt Ada. He soon met Lucy May Cole, and they were married November 24, 1908. The couple, who lived with Ada, would have three daughters and a son. Ada died in 1914, leaving her home to her nephew, and he lived there for most of the rest of his life.

Life's Work

Throughout his career, Russell worked on a wide range of areas in astronomy, studying geophysics, the stars, planets, nebulas, and the moon. He did not, however, study much of the galaxies. Russell's earliest work on binary stars provided new ways to compute parallaxes for stars. (Parallaxes are important in determining stellar distances.) His studies of eclipsing binary stars allowed him to determine the sizes of stars, and he determined new ways to estimate the masses of binary star systems. His work, always careful and painstakingly accurate, yielded a new style of precision for researchers.

Russell discovered several critical important relationships among stars. By knowing the distances of stars, and their apparent magnitudes (how bright the stars appear in the sky), he could determine the absolute magnitudes of stars (how luminous a star is in actuality). He observed that there was a definite pattern between the luminosity of a star and its color. The color of a star is related to its temperature. Almost all of the very hot, blue stars were also very bright. However, only some of the cool, red stars were bright as well. Most of the cooler stars were dim. This led him to propose that cool stars could be separated into two groups. The brighter group, he reasoned, also must be much larger in size than the dimmer group.

Unknown to Russell, Danish astronomer Ejnar Hertzsprung had been working on very similar studies. When Russell was told of Hertzsprung's studies, he immediately embraced Hertzsprung's work and even adopted his terms "giant" and "dwarf" for these two groups of stars. Most of this data was in tabular form until Russell produced a plot in 1913 of absolute magnitudes versus spectral types. This plot was presented at meetings of the Royal Astronomical Society and the American Astronomical Society. Astronomers immediately recognized how such a graphical representation of the data could be used to better understand stars. American astronomers began to refer to the plot as the Russell diagram, but by the 1930's, astronomers worldwide began to call it the Hertzsprung-Russell diagram, noting Hertzsprung's contributions.

In his quest to understand stars, Russell spent much of his career working on a theory of stellar evolution. He observed that most stars appeared to lie along a strip that ran diagonally across the absolute magnitude versus spectral type plot from the upper left to the lower right. However, some stars seemed to lie along a strip leading from the upper left to the upper right on the diagram—this strip became known as the main sequence. The stars in the upper right of the strip, Russell reasoned, were Hertzsprung's giant stars, while stars on the main sequence were dwarf stars. Russell eventually settled upon a model in which stars began as giants and then became smaller and hotter until they appeared on the upper-left portion of the main sequence. Then, he hypothesized that the stars moved down along the main sequence, becoming cooler and dimmer.

Though many aspects of this model turned out to be incorrect, astronomers have found that stars do, indeed, begin in the upper right of the Hertzsprung-Russell diagram and move onto the main sequence. However, they do not move along the main sequence. Rather, when they begin to die, they again move off of the main sequence to the upper right portion of the diagram.

During World War I, Russell was eager to help in the war effort, so he became affiliated with George Ellery Hale's National Research Council. He initially worked on a project to improve artillery parallax and ranging. Later, he worked on aircraft navigation systems. After the war, he returned to his job as a Princeton professor.

To determine the chemical composition of stars, Russell worked with stellar spectroscopy. Like most astronomers of his day, he believed that the sun and stars were composed of materials similar to those that compose the earth. Because different stars showed different spectral lines, Russell believed they had different compositions. However, Cecilia Payne, in doing research for her Ph.D., concluded that the sun and all stars essentially had the same composition—almost entirely hydrogen and helium—which turned out to be a controversial finding. Russell strongly disagreed with Payne's theory because those elements are very rare on Earth. Under his influence, Payne backed down from her theory and even declared her work flawed. However, Russell, who was more interested in truth than in being right, was presented later with undeniable evidence that Payne's conclusions were correct. He became a wholehearted supporter of Payne's theory that stars were composed mostly of hydrogen and helium.

Significance

Russell's most enduring legacy was his work in determining relationships between the luminosity and temperature of stars. His plot of absolute magnitude versus spectral type became the basis of all future such graphical depictions—diagraming—of that data. The Hertzsprung-Russell diagram was critical to the development of a theory of stellar evolution and has become one of the most important and useful tools for stellar astronomers.

Russell also was instrumental in bringing theoretical physics into astronomy. Until his time, astronomy, particularly in the Americas, amounted to cataloging and describing what one observed through research. Russell, however, sought to *explain* observations and to understand the physics of stars.

In addition to his professional contributions, Russell brought astronomy to nonastronomers. He wrote a monthly column for *Scientific American* magazine and eventually became an associate editor for the magazine in charge of astronomical work. Also, he coauthored a textbook, lectured to students, and gave several public lectures on the topic of science and religion. Finally, Russell was one of the last astronomers to work as a "generalist" in the field, focusing his work on a wide range of topics. Later astronomers would tend to become specialists in certain subfields of astronomy.

Raymond D. Benge, Jr.

Further Reading

DeVorkin, David H. *Henry Norris Russell: Dean of American Astronomers*. Princeton, N.J.: Princeton University Press, 2000. A well-researched and thorough biography of Henry Norris Russell, a major figure in the history of astronomy. Includes an extensive bibliography.

_______. "Steps Toward the Hertzsprung-Russell Diagram." *Physics Today* 31, no. 3 (March, 1978): 32-39. A history of the development of Russell's most lasting legacy: the Hertzsprung-Russell diagram.

Talcott, Richard. "Making Sense of Stars." *Astronomy* 34, no. 4 (April, 2006): 53-56. A very short and easy to understand explanation of the Hertzsprung-Russell diagram and how it is used. Includes diagram examples.

Tenn, Joseph S. "Henry Norris Russell: The Twentieth Bruce Medalist." *Mercury* 22, no. 5 (September/October, 1993): 19-21. An informative, brief summary of Russell's professional work.

CARL SAGAN

American astrophysicist

One of the most well-known scientists of the twentieth century, Sagan had the unique ability to conduct significant astronomical and planetary research and make science interesting and accessible to the general public. He inspired the well-known catch phrase "billions and billions" after his emphatic use of the word "billions" to describe the universe and its objects.

Born: November 9, 1934; Brooklyn, New York
Died: December 20, 1996; Seattle, Washington
Also known as: Carl Edward Sagan (full name)
Areas of achievement: Astronomy, physics, education

Early Life

Born in Brooklyn, New York, Carl Sagan (SAY-guhn) knew from an early age that he wanted to be an astronomer when he grew up. At that time, however, astronomy was a somewhat obscure field, and when the eight-year-old Sagan asked a librarian for a book on stars, she gave him a book on Hollywood film stars. Sagan persisted and eventually found what he wanted. Fascinated by the fact that the stars are like Earth's sun but merely farther away, he continued to read about science on his own. He also supplemented his reading with science fiction magazines, which he carefully evaluated for their scientific accuracy.

Even while avidly pursuing these interests, Sagan did not yet know that it was possible to make a living as an astronomer—he thought he would have to get a "regular" job during the day and indulge his astronomy hobby on evenings and weekends. While attending high school in Rahway, New Jersey, where his family had moved at the end of World War II, Sagan learned that it was indeed possible to pursue a career in astronomy. Fortunately, his parents supported his ambitions, and Sagan began studying physics at the University of Chicago in 1951 at the age of sixteen.

The University of Chicago, which had several Nobel Prize-winning scientists on its faculty, was an ideal environment for Sagan. He learned that other scientific fields such as biology and chemistry were also relevant to his interests, which had expanded to include the origin of life. He even organized a series of science lectures on campus, foreshadowing his later success as a popularizer of science.

Sagan continued his studies after graduation by pursuing a doctoral degree in astronomy and astrophysics under Dutch astronomer Gerard Kuiper at the University of Chicago campus in Williams Bay, Wisconsin. Before receiving his doctorate in 1960, Sagan met and married biologist Lynn Alexander, with whom he later had a son, Dorian. Sagan held academic posts at various institutions, including Harvard University, before accepting a permanent professorship at Cornell University in Ithaca, New York, in 1968, where he remained until his death.

From the beginning of his academic career, Sagan proved to be an energetic researcher and writer, publishing numerous papers and articles over the years. In 1966, he wrote a nontechnical, educational book on planets that was published as part of a Time-Life science book series. In the same year, he coauthored *Intelligent Life in the Universe* with Soviet scientist I. S. Shklovskii, an unusual collaboration in light of the Cold War hostilities between the two nations at that time. Sagan also conducted significant research on such topics as the greenhouse effect on Venus and the composition of the atmosphere of Titan, one of Saturn's moons.

Life's Work

By this time Sagan was fairly well known in the scientific world as an expert on the possibility of extraterrestrial life. In fact, when Arthur C. Clarke and Stanley Kubrick began working on the film *2001: A Space Odyssey* (1968), they consulted with Sagan about what the alien creatures in the film should look like. Ultimately, they followed Sagan's advice and decided not to show the aliens.

Sagan's entry into the public's view, however, did not occur until 1972, when he made a brief appearance on *The Tonight Show*, hosted by comedian Johnny Carson, to promote a book of his essays titled *Cosmic Connection* (1972). Carson liked Sagan enough to invite him back a few weeks later, at which time the scientist spoke about one of his favorite topics, the history of the universe. Viewers were so enthralled with Sagan's knowledge and charisma that he became a regular guest on *The Tonight Show*. Never before had a practicing scientist become so widely known to the American public.

After his success on *The Tonight Show*, Sagan realized that the general public could indeed find science fascinating if presented to them correctly and that he had a talent for doing so without abandoning the actual research itself. The dual nature of Sagan's contributions is illustrated by the fact that he simultaneously served as editor in chief of *Icarus*, a highly technical scientific journal, and wrote several articles for *Parade*, a magazine supplement distributed in newspapers all over the United States every Sunday.

In the meantime, Sagan and his first wife had divorced, and Sagan married Linda Salzman, an artist, with whom he had two sons, Jeremy and Nicholas. In the early 1970's, Sagan and Salzman worked together with the National Aeronautics and Space Administration (NASA) on the Pioneer project. Pioneer 10 was a NASA spacecraft designed to explore the outer solar system in general and the planet Jupiter in particular before becoming the first human-made object to leave the solar system. Inspired by his interest in the possibility of alien life, Sagan came up with the idea of placing a message from humankind on the spacecraft in case it ever encountered an alien race, even though that was unlikely to happen for thousands or millions of years, if ever. NASA agreed, and Sagan began designing the message, which was placed on a gold anodized aluminum plate on the spacecraft. The basic message employed mathematic and scientific concepts, which Sagan believed made up a "universal language" that could be understood by any intelligent race. Salzman contributed a drawing of two nude human figures.

Sagan's interstellar message received much public attention, and, in spite of some negative reactions to the nudity of the male and female figures, most people supported the idea of sending a message to the stars. In fact, the project received enough support that it was repeated on the Pioneer 11 spacecraft as well as on the two Voyager spacecraft launched in 1977. Ironically, Sagan ultimately collaborated with his third wife, Ann Druyan, on the Voyager message; Druyan's brain waves were recorded on a gold-plated phonograph record, along with greetings in dozens of languages, that was included on the Voyager spacecraft. Sagan and Druyan married in 1981, one month after his divorce from Salzman became final, and they remained married until the scientist's death in 1996. Sagan and Druyan had a daughter, Alexandra (called Sasha), and a son, Sam.

Even during the busiest periods of his life, Sagan did not neglect his writing. In 1977 he published *The Dragons of Eden: Speculation on Human Intelligence*, which spent several months on the best-seller list of *The New York Times* and won the Pulitzer Prize for general nonfiction. He also turned his efforts to a new medium, television. Dismayed by the relative lack of media attention given to the various NASA projects, Sagan conceived of the idea of a regular television series about the solar system and the universe. A public television station in Los Angeles agreed to sponsor a thirteen-part series called *Cosmos*, which became Sagan's

Greetings from Earth

A committee chaired by Carl Sagan selected the contents of a twelve-inch gold-plated copper phonograph record, known as the Voyager Golden Record, which was placed on each of the Voyager spacecraft to carry information about the life and cultures on Earth to any form of extraterrestrial life the craft might encounter. The record included more than one hundred images, various natural sounds, and musical selections from different periods and cultures as well as spoken greetings in fifty-five languages. U.S. president Jimmy Carter contributed the following statement, recorded on July 29, 1977.

This Voyager spacecraft was constructed by the United States of America. We are a community of 240 million human beings among the more than 4 billion who inhabit the planet Earth. We human beings are still divided into nation-states, but these states are rapidly becoming a single global civilization.

We cast this message into the cosmos. It is likely to survive a billion years into our future, when our civilization is profoundly altered and the surface of the Earth may be vastly changed. Of the 200 billion stars in the Milky Way galaxy, some—perhaps many—may have inhabited planets and spacefaring civilizations. If one such civilization intercepts Voyager and can understand these recorded contents, here is our message:

> This is a present from a small distant world, a token of our sounds, our science, our images, our music, our thoughts, and our feelings. We are attempting to survive our time so we may live into yours. We hope someday, having solved the problems we face, to join a community of galactic civilizations. This record represents our hope and our determination, and our goodwill in a vast and awesome universe.

most famous work. Although the project employed well over one hundred people, including Druyan as cowriter, Sagan himself narrated the series and, rather unusually, appeared on camera most of the time. The series, which was filmed at locations all over the world, took three years to produce and was aired in 1980. *Cosmos* won three Emmy Awards and was eventually seen by over 500 million viewers in sixty different countries. In addition, an accompanying book by the same title, also published in 1980, became the best-selling nonfiction science book in U.S. history up until that time.

Sagan also took on another role: public policy activist and adviser. After studying the effect of massive dust storms on Mars, Sagan began to calculate how vast amounts of dust could affect Earth's climate. In 1983, a controversial television film, *The Day After*, gave a fictional account of the effects of nuclear war. Immediately after the show, a panel of scientists that included Sagan discussed these effects, and Sagan voiced his opinion that a "nuclear winter," brought about by debris from nuclear bombs blocking the sunlight, could completely devastate the Earth's climate and possibly cause humankind's extinction. While many scientists disagreed with Sagan's belief that the effects would be so devastating, enough people paid attention to the theory that it prompted some historians to claim that Sagan played a significant role in nuclear disarmament and the end of the Cold War.

Finally, to reach yet another segment of the public, Sagan published *Contact*, his first and only science fiction novel, in 1985. Because of his enormous popularity, Sagan was paid a $2 million advance for the book, an unprecedented amount for a first novel. *Contact*, a story about the first message from an extraterrestrial race, received mixed but generally positive reviews. Sagan had originally intended *Contact* to be produced as a motion picture, but after many delays he decided to turn it into a novel first. A film version starring Academy Award-winning actor Jodie Foster was finally released in 1997. Unfortunately, although Sagan worked on the film production up until the time of his death, he did not live to see its actual release. In 1995, Sagan had been diagnosed with myelodysplasia, a rare bone-marrow disorder that can lead to leukemia. Several bone-marrow transplants extended his life, but on December 20, 1996, he died of pneumonia, a complication caused by the disease.

Significance

Sagan was simultaneously a scientist and a celebrity whose enormous popularity both helped and hurt his career. Although almost all scientists recognize that public awareness and support are essential to scientific research, scientists have historically shunned any colleague who receives media attention. From his early college days when he organized public science lectures at the University of Chicago, Sagan was criticized for "calling too much attention to himself" and for being too undignified for a scientist. In a commentary titled "Kinship with the Stars" in *Discover* in 1997, Jared Diamond asserted that Sagan was rejected for membership in the National Academy of Sciences—even though he had already been provisionally approved—because of his fame. The same scientists who continually bemoaned the public's lack of interest in science felt that Sagan could not possibly be a serious scientist because he spent too much time in the public eye. Eventually the academy recognized its mistake and awarded Sagan its highest honor, the Public Welfare Medal in 1994. However, their earlier rejection of Sagan may have made young scientists wary of jeopardizing their scientific careers by becoming involved in public education.

In spite of the negative attitudes directed toward him by other scientists, Sagan was very successful in the wide variety of activities he pursued. He has been credited with inspiring many young people to choose science as a career path.

Amy Sisson

Further Reading

Cohen, Daniel. *Carl Sagan: Superstar Scientist*. New York: Dodd, Mead, 1987. Cohen's biography of Sagan was written for young adult readers. The book's main focus is on Sagan's unique position as a popularizer of science. Includes a bibliography and index.

Davidson, Keay. *Carl Sagan: A Life*. New York: J. Wiley, 1999. Davidson presents a well-researched account of the life and career of the man whom he describes as a true visionary.

Ginenthal, Charles. *Carl Sagan and Immanuel Velikovsky*. New York: New Falcon, 1995. Ginenthal presents an overview of a highly publicized debate between Sagan and Russian-born scholar Immanuel Velikovsky, who attempted to prove that biblical events could be explained scientifically. This book is heavily biased in favor of Velikovsky but may be of interest regarding Sagan's views on pseudoscience.

Head, Tom, ed. *Conversations with Carl Sagan*. Jackson: University Press of Mississippi, 2006. Contains sixteen interviews with Sagan conducted over a twenty-six-year period, in which Sagan discusses topics related to science, religion, and life.

Poundstone, William. *Carl Sagan: A Life in the Cosmos*. New York: Henry Holt, 1999. A detailed chronology of Sagan's life and work.

Sagan, Carl. *Cosmos*. New York: Random House, 1980. The companion to Sagan's world-renowned television series, this 365-page book contains numerous color photographs and illustrations, an introduction that discusses the origin and development of the television series, and an extensive list for further reading organized in the same topic order as the book's chapters.

_______. "Interview." *Psychology Today* 29 (January/February, 1996): 30. This informative and readable interview with Sagan touches on many of the topics he was most passionate about: science, technology, intelligence, religion, overpopulation, and nuclear winter.

Terzian, Yervant, ed. *Carl Sagan's Universe*. Cambridge, Mass.: Cambridge University Press, 1997. This book contains a collection of essays by various scientists and writers on the many areas of scientific inquiry influenced by Sagan. Published shortly after Sagan's death, the book serves as a tribute by his peers.

CLYDE WILLIAM TOMBAUGH

American astronomer

After a systematic search based on the predictions of earlier astronomers, Tombaugh discovered Pluto in 1930. He also discovered several star clusters and galaxies, studied the distribution of extragalactic nebulas, searched for small natural Earth satellites, and made observations of the surfaces of several planets and of the Moon.

Born: February 4, 1906; Streator, Illinois
Died: January 17, 1997; Las Cruces, New Mexico
Area of achievement: Astronomy

Early Life

Clyde William Tombaugh (TOM-baw), the first of six children in his family, was born on a farm in Streator, Illinois, some seventy-five miles southwest of Chicago. His father, Muron, studied mechanical engineering at the University of Illinois but did not earn a degree. He took up farming on the family homestead and married Adella Chritton in 1905. On the farm, Muron maintained a well-equipped shop of hand-operated tools and encouraged his children to use them.

Young Clyde attended a two-room schoolhouse near the family farm. A fifth-grade teacher, Susie Szabo, encouraged his interests in history and geography, leading him to wonder about landscapes on other planets. High school was farther away, requiring a seven-mile bicycle ride each way. During the winter, he lived in Streator with his paternal grandmother and two "double" cousins, whose father, Muron's brother Lee, had married Adella Chritton's sister.

In high school, Tombaugh became a voracious reader, completing the entire Bible and devoting many hours to encyclopedias and his father's books on mathematics and physics. His uncle Lee gave him two astronomy books, one on eclipses and the other on the solar system. His interest in astronomy was further stimulated in 1920 when his father and uncle bought a refracting telescope with a 2.3-inch lens and a power of 45. This eventually became Tombaugh's first telescope, which he later installed as a finder on a 16-inch telescope.

A poor corn crop in 1921 led the Tombaughs to move to a farm near Burdett, Kansas, where they took up wheat farming. After the harvest of 1925, Tombaugh began the tedious process of grinding and polishing a mirror for an 8-inch reflecting telescope. After sending the mirror away for silvering, he was disappointed to learn that it had a poor figure and would not work well. He then built a testing chamber for mirrors in a family tornado cellar and began working on a 7-inch mirror for his uncle with much better results. In 1928, he completed a 9-inch reflecting telescope for himself and began serious observing.

Tombaugh's plans to attend Kansas State University in the fall of 1928 were ended by a summer hailstorm that destroyed the crops. In December he mailed his best astronomical sketches to Lowell Observatory in Flagstaff, Arizona, for some advice. A quick and unexpected response brought an inquiry about his training and abilities for an observing position to operate a new photographic telescope for a long-exposure survey of the sky. Director Vesto Slipher wanted a dedicated amateur astronomer willing to work long hours at low wages. In January of 1929, Tombaugh departed by rail for a ninety-day trial period at Lowell and the beginning of a new career.

Life's Work

Tombaugh's most important achievement was the discovery in 1930 of Pluto, first considered a planet but redefined—not without controversy—in 2006 as a dwarf or minor planet. During this discovery, he had been following up on predictions nearly thirty years earlier by astronomer Percival Lowell. Pluto was the last of the three planets discovered in modern times with the telescope. The planet Neptune was discovered in 1846 after small deviations in the orbit of Uranus (discovered in 1781) had led to predictions of a new and more distant planet. Since the orbital motion of Uranus did not appear to be totally explained by the presence of Neptune, Lowell, in 1905, calculated the position of an unknown object he called "Planet X" that might cause the presumed orbital discrepancies. He then initiated a series of three unsuccessful searches at his observatory in Flagstaff, Arizona.

Lowell died in 1916, and further searches were delayed when his widow challenged his will, which had designated funds for the Lowell Observatory. After the suit was settled one decade later, a new 13-inch refractor telescope and wide-angle camera were built. Late in 1928, Tombaugh was hired to conduct the fourth search for Planet X. His first duties after arriving in Flagstaff on January 15, 1929, included painting the new telescope, stoking the furnace, removing snow from the observatory domes, giving tours for visitors, and training for the planet search that began in earnest on April 26. Each 14-inch x 17-inch photographic plate required about one hour-long exposure to capture a 12-degree x 14-degree region of the sky. Several plates could be exposed during a typical clear, moonless evening.

For two months, Tombaugh operated the 13-inch telescope and processed about one hundred plates, searching in the constellation Gemini as predicted by Lowell. The senior astronomers, brothers Earl and Vesto Slipher, examined the plates containing up to 300,000 stars each. They used a device called a "blink comparator" to try to detect any moving objects that might qualify as Planet X. For each small part of the sky, two identical photos were taken on two different days. By focusing the two plates on a screen in rapid alternation, the stars would remain in place, but any moving object would appear to shift back and forth. Several such moving objects appeared on each plate, but most had the faster motions of asteroids. In June, 1929, the Sliphers tired of this work and asked Tombaugh to take over the blink comparisons along with the telescope operations.

After a couple months of frustrating work, it occurred to Tombaugh that a new planet could best be identified near its opposition point (180 degrees from the Sun). At this point, the outer planets appear to move backward in a western retrograde motion as Earth passes them, opposite

Clyde William Tombaugh. (NARA)

from their usual eastward movement relative to the stars. The farther away a planet is, the slower it moves: An asteroid image would move about 7 millimeters in one day, while a planet beyond Neptune would move only about .5 millimeter per day. Now Tombaugh began to look at regions near the opposition point rather than concentrating on the predicted positions that Lowell had calculated. He could now eliminate most of the shifting images in the blink comparator and look for short retrograde tracks.

By January of 1930, the predicted region in Gemini was in opposition and could be rephotographed with better chances of success. On February 18, Tombaugh placed two plates from Gemini, one taken on January 23 and the other on January 29, into the comparator and observed a shifting retrograde image slow enough to be beyond Neptune. On February 19, he photographed the region again and found the object a bit west of its last position, where it was supposed to be after three weeks of motion, some 6 degrees from Lowell's prediction.

On February 20, the object was observed visually with the 24-inch refractor, but no hint of a disk was revealed, raising questions about whether it was big enough to affect the motions of Uranus and Neptune as calculated by Lowell. Further observations showed that the orbit was beyond Neptune, and the announcement of the planet's discovery was made on March 13, 1930, on the seventy-fifth anniversary of Lowell's birth and the 149th anniversary of the discovery of Uranus. Six weeks later, on May 1, Vesto Slipher proposed the name Pluto, the Greek god of the nether darkness, and a superposed *P* on *L* was chosen as its symbol, matching the initials of Percival Lowell.

For the next two years, Tombaugh continued his search program to be sure that no other planets might be missed. On June 1, 1932, he made his second major discovery when he identified a globular cluster of stars in our galaxy's central hub. In the fall of 1932, Tombaugh was given a leave from the observatory to take up a scholarship at the University of Kansas, where he met and married Patricia Edson in 1934 and completed an astronomy degree in 1936. Returning to Flagstaff, he continued his sky survey until 1943, interrupted only by his 1938-1939 master's degree work back in Kansas. In 1937, he discovered a supercluster of galaxies and measured its shape and size, eventually cataloging nearly thirty thousand galaxies. He found five open-star clusters and identified a total of 3,969 asteroids, of which more than seven hundred had not been observed before.

In 1943, Tombaugh began teaching physics and navigation for the U.S. Navy at Arizona State Teachers' College in Flagstaff. In 1946, he moved to Las Cruces, New Mexico, to take a position as an optical physicist and astronomer at the White Sands Proving Grounds, designing optical equipment and supervising the tracking of V-2 missile firings. In 1953, he proposed a project and received military funding to search for small natural Earth satellites; for two years he divided his time between White Sands and Lowell Observatory, where the search began.

In 1955, Tombaugh transferred the natural satellite search project to New Mexico State University to facilitate observations from Quito, Ecuador. No such satellites were confirmed, but in 1957, Tombaugh made some of the first photos of Sputnik 1, the first artificial satellite, and provided the observational basis for the safety of an extensive artificial satellite program. From 1961 to 1973, he was a professor in the Department of Earth Sciences and Astronomy at New Mexico State University, where he established a comprehensive observational program of the five nearest planets. After his retirement, he remained active in astronomy and continued to build telescopes. Over his lifetime, he ground thirty-six optical surfaces with diameters from 1.5 to 20 inches. Telescopes on which he worked are featured at the Tombaugh Observatories at both New Mexico State University and the University of Kansas.

Significance

In many ways, Tombaugh was like his astronomer hero William Herschel, who discovered Uranus unexpectedly during a routine sky survey in 1781. Both were dedicated amateur astronomers and skilled telescope makers who devoted hours to tedious observations. Tombaugh, however, was only twenty-four years old when he discovered Pluto, while Herschel was nearly twenty years older. Furthermore, Tombaugh's year-long search for Planet X lasted much longer than that of either Herschel or Johann Galle, who discovered Neptune in 1846 on the first night he looked for it, lying less than 1 degree from its predicted position. The search for Pluto was complicated because its orbit is highly eccentric—sometimes even passing inside the orbit of Neptune—and has a large inclination of about 17 degrees from the mean plane of the other planets.

It is now known that Lowell's predictions for the position of Pluto were based on faulty calculations, and its discovery within 6 degrees of the predicted location was only a coincidence. Fortunately, Tombaugh did not limit his observations to the predicted area of the sky or to the region close to the mean orbital plane of the planets. When James Christy discovered Pluto's moon, Charon, in 1978, it was conclusively demonstrated that the mass of Pluto was far too small to cause observable deviations in the orbits of Uranus and Neptune; thus the two larger planets' orbits could not be used to predict Pluto's position. In the 1990's, several icy objects much smaller than Pluto were discovered just beyond its orbit in the Kuiper comet belt with periods of about three hundred years, compared to Pluto's 248-year period.

Joseph L. Spradley

Further Reading

Adler, Larry, Mary Carmichael, Nomi Morris, and A. Christian Jean. "Of Cosmic Proportions." *Newsweek* 148, no. 10 (September 24, 2006): 44-50. Examines how and why Pluto lost its planetary status and how that decision "changed" the shape of the solar system. Includes discussion of Tombaugh's discovery of Pluto.

Kaufmann, William J. *Universe*. 4th ed. New York: W. H. Freeman, 1994. This textbook is a good standard introduction to astronomy with accurate descriptions of the discovery of Pluto, its properties, and its moon Charon.

Levy, David H. *Clyde Tombaugh: Discoverer of Planet Pluto*. Tucson: University of Arizona Press, 1991. A complete biography, written by a fellow astronomer and based on thorough research and personal interviews.

Rath, Ida. *Boy Planet Seeker*. Dodge City, Kans.: Rollie Jack, 1963. This is a short (108 pages), illustrated children's book about Tombaugh's life (with an emphasis on his Kansas roots), his planet search, and his discovery of Pluto.

Tombaugh, C. "The Discovery of Pluto." In *Source Book in Astronomy, 1900-1950*, edited by Harlow Shapley. Cambridge, Mass.: Harvard University Press, 1960. This excerpt is an accurate and readable first-person account of the discovery of Pluto.

Tombaugh, Clyde W., and Patrick Moore. *Out of the Darkness: The Planet Pluto*. Harrisburg, Pa.: Stackpole Books, 1980. An extended account of the discovery of Pluto written for the fiftieth anniversary of its discovery, with background information on asteroids, Uranus, and Neptune by astronomy writer Patrick Moore.

Weintraub, David A. *Is Pluto a Planet? A Historical Journey Through the Solar System*. Princeton, N.J.: Princeton University Press, 2007. Weintraub examines how the definition of a planet has changed throughout history and chronicles how Pluto's claim to planetary status has been challenged since Tombaugh discovered it in 1930.

JAMES VAN ALLEN

American physicist

Van Allen pioneered the use of artificial satellites for Earth studies. His discovery of radiation belts around the earth initiated the field of magnetospheric physics. He applied his expertise to help deploy planetary probes to enhance space exploration and knowledge. A critic of piloted space missions, he urged aerospace administrators to utilize robotic spacecraft to explore and transmit information from space.

Born: September 7, 1914; Mount Pleasant, Iowa
Died: August 9, 2006; Iowa City, Iowa
Also known as: James Alfred Van Allen (full name)
Areas of achievement: Physics, astronomy, aviation and space exploration, invention and technology

Early Life

James Van Allen was born to lawyer Alfred Morris Van Allen and schoolteacher Alma E. Olney Van Allen in Mount Pleasant, Iowa. His parents encouraged their children, all boys, to become well educated, to develop a work ethic, and to be frugal. To earn money Van Allen raised chickens, gardened, and delivered newspapers. He excelled at school, particularly in mathematics and science, and studied Latin and woodworking.

Van Allen developed his innate curiosity by reading *Popular Science* and *Popular Mechanics* magazines. He constructed motors, a crystal radio, and a Tesla coil. His science teacher, Clarence A. Cottrell, provided Van Allen with intriguing and challenging physics problems, which sometimes required hours to solve. Learning about planets, Van Allen became interested in astronomy and space, speaking to classmates and his family about the possibilities of future travels to the moon. He was editor of the Mount Pleasant High School yearbook, *The Target*, and graduated as class valedictorian in June, 1931.

Van Allen enrolled in Iowa Wesleyan College, located in his hometown, and majored in physics. He came to admire his future mentor Thomas C. Poulter, who urged his pupils to pursue innovative research. Poulter taught Van Allen shop skills that would prove useful for making scientific instruments, and he later hired Van Allen as an assistant, paying him a thirty-five-cent hourly wage. Supplied with a magnetometer by the Carnegie Institution's Department of Terrestrial Magnetism (DTM), Van Allen gathered data for a field magnetic survey of Henry County, Iowa. Admiral Richard E. Byrd named Poulter chief scientist for the Second Byrd Antarctic Expedition in the early 1930's, and Van Allen was assigned to be Poulter's aide on the project, gathering instruments and equipment. In June, 1935, Van Allen received a bachelor of science degree, summa cum laude, listening to Byrd, who was commencement speaker.

Life's Work

Interested in pursuing a scientific career, Van Allen expanded his physics knowledge by seeking graduate degrees at the University of Iowa in Iowa City. He worked with adviser Edward P. T. Tyndall, earning a master of

James Van Allen. (Library of Congress)

science degree in 1936. Van Allen studied with his major professor, Alexander Ellett, to complete a doctorate in nuclear physics three years later.

Van Allen accepted a DTM position as a Carnegie Research Fellow in a nuclear physics laboratory. After World War II began, Van Allen devised radio proximity fuses for the U.S. Navy to use for antiaircraft tactics. By April, 1942, he conducted that work in the new Applied Physics Laboratory (APL) at Johns Hopkins University. In November, 1942, Van Allen was commissioned as a lieutenant (junior grade) in the Naval Reserve to serve in the South Pacific as an evaluator of the naval deployment of radio proximity fuses. Van Allen married Abigail Fithian Halsey on October 13, 1945, at Southampton, New York, and they later had three daughters and two sons.

After the war, Van Allen returned to APL to establish and chair the High Altitude Research group. The group used German V-2 rockets at White Sands, New Mexico, including a launch with Van Allen's instruments, to assess cosmic rays. By 1946, Van Allen oversaw the use of Aerobee rockets, which APL research supported instead of V-2s, for high altitude research. Both *The New York Times* and *Time* magazine criticized Van Allen for suggesting satellites could collect scientific data in the future. The National Aeronautics and Space Administration (NASA) and the National Academy of Sciences (NAS) Space Science Board began appropriating high altitude research, and APL administrative changes led to a shift in research focus. Van Allen contemplated how to continue his research.

Van Allen's former adviser, Tyndall, suggested he apply for a job as Physics Department chair at Iowa, a position he would start in January, 1951. At Iowa, Van Allen used balloons to transport instruments aloft to investigate cosmic rays. He innovated "rockoons," his term for a balloon-rocket device to launch instruments to high altitudes above Earth, advising students who participated in that research to earn graduate degrees. During the 1950's, Van Allen launched rockoons on several expeditions near Greenland and the South Pole to measure cosmic radiation.

Van Allen served as chair of a group for satellite instrumentation during the International Geophysical Year (IGY) from July, 1957, to December, 1958. Aware of military efforts to deploy an Earth satellite, he developed instrumentation compatible with both Navy and Army satellites in development. In October, 1957, while he was on a rockoon expedition in the Pacific, Van Allen heard the signal of the Soviet Sputnik satellite on the radio. Excited by Sputnik, Van Allen was eager for a U.S. satellite to orbit the earth. After the Navy's Vanguard launch failed, Van Allen's Geiger counter was placed instead on the Army's Explorer I satellite, which launched on January 31, 1958.

Assessing information collected from Explorer I, Van Allen noticed first an absence then a surge of data, concluding that radiation was affecting the measurements. He gathered information from instruments on Explorer III and IV, determining that two "radiation belts" existed around the earth. On May 1, Van Allen publicly announced his detection of the radiation belts, which were later named in his honor, in a paper at an American Physical Society and National Academy of Sciences meeting. In March, 1959, *Scientific American* printed his article describing the radiation belts. Van Allen's work inspired the field of magnetospheric physics, which intrigued researchers worldwide.

Starting in 1958, Van Allen chaired a space scientists' group that supported a piloted lunar landing, but he eventually decided robotic missions were preferable for space exploration because he thought few significant scientific results had been achieved by human space travelers. Van Allen devised instrumentation for the Mariner spacecraft deployed to Venus and Mars. Next, as principal investigator, he developed instruments for Pioneers 10 and 11, detecting and surveying radiation belts associated with Jupiter and Saturn in the 1970's. He later supported deploying the Voyager probes, which expanded exploration to the outer planets, and the Galileo project that orbited Jupiter.

Although Van Allen retired in 1985, he remained active in his department and profession. Mount Pleasant citizens restored Van Allen's boyhood home during the late 1990's, and he participated in its dedication as a historical site. Although science, not fame, motivated Van Allen, he received numerous recognitions for his accomplishments. He was awarded the Royal Astronomical Society Gold Medal, the Royal Swedish Academy of Sciences Crafoord Prize, the National Air and Space Museum Trophy, and U.S. National Medal of Science.

On October 9, 2004, the University of Iowa hosted a ninetieth-birthday celebration for Van Allen, who continued researching and evaluating satellite data in his office in Van Allen Hall. His heart failed on August 9, 2006, at University of Iowa Hospital. The university sponsored a public memorial and started a scholarship in his honor.

Significance

Van Allen shaped space science as it emerged in the mid-twentieth century. His contributions revealed new perceptions of the universe—providing crucial information to enable spacecraft to maneuver in the solar system—and enhanced perspectives of the earth. His work highlighted the potentials of space and influenced students to seek careers in space science. Many of these individuals would become prominent NASA investigators and administrators. His opinions influenced NASA policy, especially the increase of unpiloted space missions during the 1990's.

Many historians consider Van Allen's detection of the earth's radiation belts to have strengthened the position of the United States in the Cold War space race, but he conducted investigations to improve humanity and not to advance political ambitions. He nurtured a sense of scientific community in the United States, and he worked with former German rocket scientists and traveled to the Soviet Union to share insights into advancing science and technology.

Elizabeth D. Schafer

Further Reading

Beatty, J. Kelly, Carolyn Collins Petersen, and Andrew Chaikin, eds. *The New Solar System*. 4th ed. Cambridge, Mass.: Sky, 1999. Includes a chapter written by Van Allen and astrophysicist Frances Bagenal describing their work with magnetospheres. The chapter incorporates examples from throughout Van Allen's career, including planetary exploration with probes. Provides numerous charts and illustrations including Hubble space telescope images.

Bille, Matt, and Erika Lishock. *The First Space Race: Launching the World's First Satellites*. Foreword by James Van Allen. College Station: Texas A&M University Press, 2004. Provides biographical details on Van Allen and his research. Van Allen praised this text for providing historical context for U.S. and Soviet aerospace research conducted in the 1950's.

Foerstner, Abigail. *James Van Allen: The First Eight Billion Miles*. Iowa City: University of Iowa Press, 2007. Comprehensive biography written by a Northwestern University journalist who interviewed Van Allen and consulted his archival papers, chronicling his scientific curiosity, philosophy of space research, and achievements from childhood through his radiation discoveries with satellites and probes.

Newell, Homer E. *Beyond the Atmosphere: Early Years of Space Science*. Washington, D.C.: National Aeronautics and Space Administration, 1980. Newell, a physicist and NASA administrator affiliated with White Sands, Vanguard, and IGY research notes Van Allen's leadership in pioneering space-science projects and groups. Appendixes, tables, figures.

O'Brien, Brian. "Radiation Belts." *Scientific American* 208 (May, 1963): 84-96. Although written five years after the discovery of the Van Allen radiation belts, and somewhat dated, this article provides a wealth of information about them, including maps of the belts around the earth and the shape of the earth's magnetic field.

Van Allen, James A. *Origins of Magnetospheric Physics*. Iowa City: University of Iowa Press, 2004. Detailed account of Van Allen's research and experiences, including hearing the radio signals of Sputnik. Includes images from his field notebooks, radiogram and correspondence transcripts, diagrams, and research proposal excerpts.

WANG ZHENYI

Chinese astronomer, mathematician, and poet

Wang Zhenyi's promotion of the Western, Sun-based calendar, as well as her research, experiments, and observations, contributed to China's modern understanding of astronomical phenomena such as lunar and solar eclipses, equinoxes, and celestial mechanics. Her mathematics textbooks were written for broad access, and her poetry was praised for its strength, clarity of vision, and social compassion.

Born: 1768; Nanjing, China
Died: 1797; Nanjing, China
Also known as: Wang Chen-i (Wade-Giles)
Areas of achievement: Astronomy, mathematics, literature

Early Life

Wang Zhenyi (wang jehn-YEE) was born to a scholarly, aristocratic Manchu family in the large town of Nanjing, China. Originally from the neighboring province of Anhui to the west, her well-educated grandfather, Wang Zhefu, had worked as county governor and prefect in the Nanjing region. Her father, Wang Xichen, had failed the imperial examination that would have set him on a course similar to that of an administrator. Instead, he became a physician, publishing the results of his clinical studies. No personal details have survived about Wang's mother, an upper-class Manchu woman.

When Wang was a child, her grandfather and father encouraged her innate thirst for knowledge and allowed her to learn to read and peruse the first volumes of her grandfather's seventy-five bookcases. Supporting a girl's formal education was unusual for a Chinese family of the time, and it gave Wang uncommon access to the world of learning. She would prove to be exceptionally intelligent. As a young Manchu woman, Wang was spared the torturous practice of foot-binding inflicted on upper-class Han Chinese women.

Wang's grandfather died in 1779, when Wang was eleven years old. She traveled with her paternal grandmother Dong and her father to Jilin in Manchuria, where her grandfather was given an ornate funeral. While in Jilin, Wang enjoyed unusual access to books and learning as well as physical activities. With three upper-class girlfriends, and under the tutelage of an older woman, Wang continued to read voraciously, and she quickly gained a superlative knowledge of literature and the natural sciences. She learned horse-riding and arrow-shooting (horseback-based martial arts) from a Mongolian general who took a liking to his unlikely female disciple. Wang became an expert markswoman and was rumored to have hit the target with every arrow shot while galloping on her horse.

In 1784 sixteen-year-old Wang returned south with her father, traversing Shaanxi province, crossing through Hubei province, and venturing south into Guangdong (Canton) before settling again in Nanjing. Wang gained a wide view of the Qing Dynasty, an opportunity afforded very few women of her generation.

At the age of twenty-five, unusually late for the time period, Wang married Zhan Mei, who was from her ancestral Anhui province. Their marriage is reported to have been happy, but they had no children.

Life's Work

Wang Zhenyi's contributions to astronomy were groundbreaking. Her treatise on the observation and calculation of the vernal and autumnal equinoxes—the one day in spring and fall, respectively, when there are exactly twelve hours each of day and night—accurately described the causes of the phenomenon and how to calculate its exact day each year.

Her work on lunar and solar eclipses was uncharacteristically based on an actual experiment, something very few of her Chinese male colleagues had ever thought of doing. To study the mechanics of an eclipse, when either the Sun covers the Moon at night or the Moon occludes the Sun during the day, Wang built her own scientific apparatus in a pavilion of her garden in Nanjing. She used a round table to represent the Earth; above the table, she attached a crystal lamp functioning as the Sun; a large round mirror on the side of the table (Earth) stood in for the Moon. Moving table, lamp, and mirror according to the real movement of Earth, Sun, and Moon, Wang could observe in her own pavilion the actual effects causing the eclipses of Sun and Moon. Her written work on the subject proved exceptionally accurate and furthered Chinese astronomical understanding.

Wang also affirmed the modern view of the Earth as a sphere and wrote on celestial mechanics affecting the movement of the Sun and the planets. Her meteorological studies concerned calculating the humidity in the atmosphere to arrive at better weather forecasts. Wang saw her work regarding the meteorological prediction of floods and droughts as a service to the farmers of China, who suffered much from these extreme weather conditions.

In the field of mathematics Wang showed a clear understanding of trigonometry, demonstrating an innate skill for abstract thought. Her popularizing textbooks on calculations, first published when she was twenty-four years old, were designed to enable beginners to gain a quicker understanding of multiplication and division. Her work demonstrated a clear grasp of the principles of mathematics, and she conveyed this clarity to her readers. She also wrote on problems in advanced mathematics that were at the cutting edge of the science of her day.

Wang Zhenyi enthusiastically embraced the Western Sun-based, or heliocentric, calendar. She praised the underlying principles that gave it accuracy and precision, appreciated its calculations, and she argued for its adoption in China. She opposed keeping the older, less accurate lunar calendar for the sake of tradition, and she argued for acceptance of new scientific and mathematical results and ideas.

Wang Zhenyi was also an accomplished poet whose work showed concern with the experiences of women and common people, and with her yearning for knowledge, traveling, and freedom. One of her earliest poems, which describes her journey to Jilin for her grandfather's funeral, has been praised for its perception and insight. Often her poetic persona is that of a strong, adventurous woman who views, for example, the world from the perspective of the gods: looking down a mountain pass to the Huang He (Yellow River) in the plains below.

Wang's poems, written during her travels with her father before her marriage, show a keen social compassion and understanding of the economic forces threatening the Qing Dynasty, which would lead to civil war in the next century. She wrote about peasant women and the callous indifference of the wealthy, who hoarded rice until it rotted, leaving many people to starve. She also criticized the ever-increasing tax burden that stifled rural development. Such political topics were unusual in women's poetry of her time.

In 1797, at age twenty-nine, Wang died of natural causes. Shortly before her death she entrusted all of her manuscripts to her best woman friend, Qian Yuling. In 1803, Qian conveyed the manuscripts to her nephew Qian Yiji, who compiled Wang's mathematics textbook *Shusuan jiancun* (eighteenth century; simple principles of calculation) and wrote a preface praising Wang's intellectual achievements. After being transferred to Zhu Xuzeng, a Nanjing collector, most of Wang's manuscripts were lost. Only her *Defengting zhuji* (eighteenth century; first collection of the Defeng Pavilion), four volumes of poetry, and the Jilin collection of essays appear to have survived. Most of her astronomical and other scientific work, in its original form, has been lost, and is thus known only through references by others.

Significance

Wang Zhenyi's accomplishments as an astronomer modernized Chinese astronomical understanding. Her experiment-based research linked Chinese astronomy to that of the West, which had advanced significantly by the eighteenth century. Wang helped close a gap in scientific knowledge.

Wang's work in meteorology was driven by a desire for the practical and beneficial applications of her weather forecasts, which led to her helping in the improvement of farming conditions. Furthermore, she preferred the scientifically superior Western calendar over the traditional Chinese model, demonstrating scientific openness and acceptance of results regardless of tradition or national origin.

Wang's literary work created a misogynist opposition. Upon the self-financed publication of her poetry shortly before her untimely death in 1797, male critics showed their sexism when they tried to ridicule Wang's accomplishments in poetry and extending into the field of science as well.

It took almost two centuries to appreciate the value of Wang's work. In 1994, in appreciation of her astronomical research, a small crater on Venus was named Wang Zhenyi.

R. C. Lutz

Further Reading

Liu, Naihe. "More than a Stargazer." In *Departed but Not Forgotten*, edited by Bai Shoyi. Beijing: China International, 1984. Illustrated review of Wang's life and her scientific and literary accomplishments. Quotes from her poetry and places her life in sociohistorical context.

Meschel, Susan V. "Teacher Keng's Heritage: A Survey of Chinese Women Scientists." *Journal of Chemical Education* 69 (September, 1992): 723-729. Illustrated survey of Wang's life and accomplishments based on Naihe Liu's study. Places Wang in the context of Chinese women scientists whose traditional cultural and social status is juxtaposed with the official policies of the People's Republic of China.

Spence, Jonathan. *The Search for Modern China*. 2d ed. New York: Norton, 1999. Most widely available book in English on modern Chinese history. Useful in providing background information on the lives Wang Zhenyi and more-conformist women led in the eighteenth century Qing Dynasty. Maps, illustrations, notes, bibliography, index.

Xu, Zhentao, et al. *East Asian Archaeoastronomy: Historical Records of Astronomical Observations of China, Japan, and Korea*. Amsterdam: Gordon & Breach Science, 2000. Places Wang's astronomical research and accomplishments in the context of Chinese astronomy. Highlights the innovative value of Wang's observations, calculations, and theories. Illustrated.

MAX WOLF

German astronomer

Wolf was the first astronomer to use an astronomical camera to discover asteroids by combining the camera with a mechanical telescope. During his very full career, Wolf discovered 582 asteroids with 228 of these receiving general recognition. This figure is a personal record of discoveries in astronomy that has been difficult to surpass.

Born: June 21, 1863; Heidelberg, Baden (now in Germany)
Died: October 3, 1932; Heidelberg, Germany
Also known as: Maximillian Franz Joseph Cornelius Wolf (full name)
Area of achievement: Astronomy

Early Life

Max Wolf (vohlf) was the son of Franz Wolf and Elise Halwerth. He was attracted to astronomy at an early age and, while a student at Heidelberg, erected his own small observatory. Since his father was a wealthy physician, Wolf was able to afford this private facility. In 1884, he discovered the comet that bears his name. This discovery so fired his interest in astronomy that he gave up all of his other studies to pursue a career in this field, earning a Ph.D. from Heidelberg with a dissertation in celestial mechanics.

In 1891, Wolf discovered his first asteroid, using photography. For this discovery, he was awarded the Lalande Prize of the Paris Academy of Sciences. After a visit to the United States, during which he received financial support for a sixteen-inch double telescope, Wolf returned to Heidelberg to work with the grand duke of Baden to build a new observatory at Königstuhl. In 1893, Wolf was named director of this new observatory and, simultaneously, became extraordinary professor of astrophysics at the University of Heidelberg. In 1902, he was elected to the chair of astronomy at the same university. Wolf remained at Heidelberg for the rest of his life, discovering his last asteroid, 1219 Britta, there only a few months prior to his death.

Life's Work

Wolf had two major accomplishments during his career. The first was the use of an astronomical camera in connection with a telescope to hunt for asteroids. To accomplish this search, the telescope was centered on a conveniently located bright star. The clockwork of the telescope kept track with the movement of the star so that photographs showed the star as a dot of light. Asteroids and comets moving across the field around the star would appear on the photograph as lines. Visual observation with the telescope alone would then determine whether the line was a comet or an asteroid. Because the clockwork drive for telescopes was not entirely accurate, the photographic session had to be monitored constantly to keep the star in the crosshairs of the telescope. Wolf developed the technique for this method of sky search to a high degree. His methods were adapted widely by other astronomers.

The second great accomplishment of Wolf's career was the discovery of the Trojan asteroids, a large number of asteroids outside the asteroid belt and in the orbit of Jupiter. In 1906, Wolf observed an asteroid, which he called Achilles, whose orbit seemed unusual. Mathematical computations showed that Achilles was in the orbit of Jupiter. Although this seems impossible, an eighteenth century mathematician, Joseph-Louis Lagrange, had determined that, in theory, a small object can travel in the same orbit as a large object if the two objects and the sun form an equilateral triangle. To do this, the small object must be about sixty degrees ahead of or behind the large one. Wolf was searching sixty degrees ahead of Jupiter when he discovered Achilles. The name "Achilles" was chosen by Wolf because other asteroids had feminine names, but this one was outside the asteroid belt.

The discovery of Achilles suggested that one should search sixty degrees behind Jupiter, and there other bodies were found. The first of these was named Patroclus. This established a pattern; all these asteroids were named for heroes of the Trojan War, and the two groups are collectively called the Trojan asteroids. All of those in the Achilles group, except one, Hector, are named for the Greeks, and, except for Patroclus, the second group are named for the defenders of Troy. Some seven hundred Trojan asteroids have been identified.

Significance

By combining the best technology of his time with accepted theories, Wolf was able to demonstrate the factuality of the Lagrange theory with respect to Jupiter. No other planets have asteroids at the Lagrange points of sixty degrees ahead or behind. By demonstrating an effective method of searching for asteroids, Wolf advanced early twentieth century knowledge of astronomy. The technology of astronomy of the late twentieth century has led to so many asteroid discoveries that one astronomer has called them "vermin of the skies." In Wolf's day, they were a new and exciting field of observation and discovery.

Michael R. Bradley

Further Reading

Aguirre, Edwin. "A Flaming Runaway Star." *Sky and Telescope* 109, no. 4 (April, 2005): 32-33. Describes IC 405, which Wolf called a flaming star.

Croswell, Ken. "The Blue Witch." *Sky and Telescope* 113, no. 4 (April, 2007): 26-27. Discusses the colors of interstellar nebulae, including the Witch Head Nebulae discovered by Wolf.

Cunningham, Clifford J. "The First Planetarium." *Mercury* 34, no. 5 (September/October, 2005): 10. A brief history of the modern planetarium, which originated in 1913 when Wolf suggested the creation of a planetarium to be installed at a museum in Munich.

Freiesleben, H.-Christ. "Maximilian Franz Joseph Cornelius Wolf." In vol. 14 of *Dictionary of Scientific Biography*, edited by Charles Coulston Gillispie. New York: Charles Scribner's Sons, 1976. This reference work has a biographical sketch of Wolf and includes a short bibliography of his astronomical articles, most of which are in German and are untranslated.

Gehrels, Tom. *Asteroids*. Tucson: University of Arizona Press, 1979. This textbook on asteroids has several references to Wolf and his work. It explains the significance of asteroids in the planetary scene.

Kowal, Charles T., ed. *Asteroids: Their Nature and Utilization*. New York: John Wiley & Sons, 1988. Kowal describes Wolf's discovery of the Trojan asteroids and discusses the Lagrange theory in layperson's terms.

Mitton, Simon, ed. *The Cambridge Encyclopedia of Astronomy*. New York: Crown Books, 1977. A basic reference work, this book has good articles on asteroids and the Trojan asteroids. Includes a brief biographical sketch on Wolf that focuses on his work.

Vaucouleurs, Gérard Henri de. *Astronomical Photography: From the Daguerrotype to the Electron Camera*. New York: Macmillan, 1961. A history of asteroid photography. Contains numerous references to Wolf's work and several photographs and charts of asteroids.

INDEX